Digital Image Processing in Remote Sensing

Digital Image Processing in Remote Sensing

Edited by
Jan-Peter Muller
Department of Photogrammetry and Surveying
University College London

Taylor & Francis
London and Philadelphia
1988

First published 1988 by Taylor & Francis Ltd
4 John Street, London WC1N 2ET
Published in the USA by Taylor & Francis Inc.
242 Cherry Street, Philadelphia, PA 19106–1906

© 1988 Taylor & Francis Ltd

All rights reserved. No part of publication may be produced, stored in a retrieval system or transmitted, in any form or by any means, electronic, mechanical, photocopying or otherwise, without the prior permission of the Copyright owner.

Typeset by John Kolbusz using Donald Knuth's TeX language.
Printed in the United Kingdom by
Taylor & Francis (Printers) Ltd.
Rankine Road, Basingstoke, Hampshire RG24 0PR

British Library Cataloguing in Publication Data
Main entry under title:

Muller, Jan-Peter
  Digital image processing in remote sensing.
  1. Remote sensing. Applications of
  digital image processing
  I. Title
  621.36'78'0285

ISBN 0-85066-314-8

Front cover photograph

This perspective view was produced by overlaying a 20m SPOT multispectral image on an automatically stereo matched DEM derived from SPOT 10m data. It is representative of the new ways that remotely sensed data will be exploited in the future with the advent of routine DEM information generated automatically from satellite.

To Sheila, James, Ruth and Edna

Preface

Digital image processing is now a well-established tool of remote sensing and has been employed by scientific disciplinary researchers and practioners since the early days of the Earth observation program. The breadth and scope of the subject area means that nowadays it encompasses subjects as varied as physics, mathematics, electrical and electronic engineering and, increasingly in the last ten years, computer science.

During the early 1980s as a direct result of the drastic reductions in the price of computing equipment, a number of groups around the world tackled the same problems of how to develop a tool that was sufficiently broad in scope and yet could be used by application specialists. At the same time, the techniques which had been developed for the interplanetary space-probes began to be applied to terrestrial data and vice versa. This cross-fertilisation between application disciplines is one of the hallmarks of digital image processing in remote sensing and is one of the themes which emerges from this text.

In 1984, a joint workshop, sponsored by IBM UK Limited, was convened between the Remote Sensing Society and the newly created Centre for Remote Sensing at Imperial College. Over 200 delegates came to the meeting and heard some sixteen presentations from speakers from France, United States and the UK. This book contains seven chapters from speakers at that meeting as well as other experts from the field.

The aim of that workshop and one of the primary themes of this book was to examine the computing issues involved in image processing from the point of view of developments in computer science largely stemming from the increased availability of powerful computing equipment with virtual memory management. The impact of these developments on various application disciplines from geology to weather forecasting has been profound and many of the analyses described here would have been inconceivable otherwise.

This book is therefore aimed at those who would like to discover how image processing is done including with what equipment and software and using which techniques. There are many texts which give detailed mathematical treatment of the fundamental principles behind a particular technique. However, for many scientists these texts are unhelpful.

Issues addressed in this book include Human-Computer Interfaces for both research and operational applications of digital image processing from the standardised flexible user interface developed by the NASA Goddard Space Flight Centre for application scientists to an extension of the electronic paintbox for the production of television satellite movies to a language for the prototyping of new techniques. Hardware systems play a key role in any image processing work and although hardware changes almost with the seasons in our technological helter-skelter, the demands from users for cheap standalone systems is stronger now than even a few years ago.

The merging of raster and vector data is a key issue in the integration of digitised map data with satellite images and the merging of features extracted from different satellite sensors in one common geographical format. The chapter by Bell and colleagues addresses this question head on with a novel approach to the uniform manipulation of these multifarious data types.

In the past, reports on applications of remote sensing have tended to be in the land use/thematic mapping area. I have therefore tried to include chapters by experts in other fields with the emphasis on how image processing can help an application scientists in areas as diverse as nowcasting, ocean dynamics, optical astronomy, geology and evapotranspiration monitoring.

I would like to express my gratitude to John Kolbusz for editorial work and page composition, to Phil Taylor at Royal Holloway & Bedford New College and Bob Colville and Claire Steward at ULCC, for typesetting this book at short notice. I also thank John Kolbusz, Tim Day, James Pearson, Sam Richards, Ananda and Mike Dalton for the computer graphics used to illustrate the cover and the first two Chapters. I am deeply grateful to Mike Barnsley for subediting and index preparation at the last moment, and acknowledge discussions with D. Clark, G. Garneau, J. Mosher, G. Yagi, M. Martin, M. Easterbrook and R. Gorley. R. Hake and G. Robinson of IBM are thanked for their support of the image processing workshop. This book would not have been possible without the continued support of my family and friends to whom I am indebted.

This book has been an experiment in book production by computer, and as such has been subject to technical pitfalls along the way. I apologise to the authors and readership for the delays that have ensued before publication.

IAX is now a registered IBM product (program number 5788-EDD).

Jan-Peter Muller
University College London
July 1988

Contents

Preface	....	vii
Chapter 1.	Computing issues in digital image processing in remote sensing	1
	J-P. Muller	
1.1.	Introduction	1
1.2.	Hardware Issues	4
1.3.	Software Issues	11
1.4.	Image Processing Workstations—an Examplar	14
1.5.	Conclusions	16
Chapter 2.	Visualisation of topographic data using video animation	21
	J-P. Muller, T. Day, J. P. Kolbusz	
	M. Dalton, S. Richards, J. C. Pearson	
2.1.	Introduction	22
2.2.	Objectives	23
2.3.	Surface representation	24
2.4.	Rendering	26
2.5.	Display	31
2.6.	Discussion and conclusions	36
Chapter 3.	The Transportable Applications Environment —an interactive design-to-production development system	39
	D. C. Perkins, D. R. Howell & M. R. Szczur	
3.1.	Overview	39
3.2.	TAE features	42
3.3.	Building a System with TAE	53
3.4.	Using a System Built with TAE	57
3.5.	TAE Utilisation	60
3.6.	Summary	62
Chapter 4.	A menu-based interface oriented to display processing of real-time satellite weather images	65
	K. Tildsley & C. England	
4.1.	Introduction	65
4.2.	Menu Interaction—The User Viewpoint	66

4.3.	Processes	70
4.4.	Menu Interaction—The Programmers View	71
4.5.	Processes	74
4.6.	Speed Of Execution	75
4.7.	Device Independence	76
4.8.	Applications Programs	77

Chapter 5.	IAX—An Algebraic Image Processing Language for Research	79
	P. H. Jackson	
5.1.	Introduction	79
5.2.	Languages Overview	80
5.3.	Design Criteria	82
5.4.	IAX Language Concepts	83
5.5.	Data Formats and Types of IAX Variables	84
5.6.	Expressions	85
5.7.	Statement Types	88
5.8.	Input and Output	91
5.9.	Some Examples	91
5.10.	Description of Functions, Commands & Pseudo-Variables	99

Chapter 6.	Microcomputers and Mass Storage Devices for Image Processing	105
	D. C. Ferns & N. P. Press	
6.1.	Overview	105
6.2.	Introduction	105
6.3.	Microcomputer Hardware Configurations	109
6.4.	Storage and Archiving Media	115
6.5.	Conclusions	120

Chapter 7.	A very low-cost Microcomputer-based Image Processor	123
	P. J. Beaven	
7.1.	Overview	123
7.2.	Introduction	123
7.3.	Development of an image processor	124
7.4.	Equipment	126
7.5.	Language	127
7.6.	Program development	128
7.7.	Data transfer	130
7.8.	Subsequent hardware development	131
7.9.	Specialist systems	131
7.10.	Conclusions	132

Chapter 8.	Capturing Image Syntax using Tesseral addressing and arithmetic	135
	S. B. M. Bell, B. M. Diaz & F. C. Holroyd	
8.1.	Introduction	135

8.2.	Tesseral Addressing and Arithmetic	136
8.3.	Quadtree Storage	136
8.4.	Tesseral addressing of Quadtrees	137
8.5.	Tesseral Raster Storage	139
8.6.	Tesseral Arithmetic	141
8.7.	Rule-based tesseral arithmetic	146
8.8.	Use of Tesseral Methods in Image Manipulation	149
8.9.	Summary and Conclusions	151
Chapter 9.	Multiple source data processing in remote sensing	153
	P. T. Nguyen & D. Ho	
9.1.	Introduction	153
9.2.	Data Correction	154
9.3.	Data Presentation and Storage	158
9.4.	Applications in Geology	161
9.5.	Applications in climatology	170
9.6.	Conclusion	174
Chapter 10.	Extreme Variability, Scaling and Fractals in Remote Sensing: Analysis and Simulation	177
	S. Lovejoy & D. Schertzer	
10.1.	Overview	177
10.2.	Introduction	178
10.3.	Analysis Techniques for Mono and MultiDimensional Phenomena	182
10.4.	A Simple Monodimensional model with strong intermittency	188
10.5.	The Dimension of Surface Network and the need for Remote Sensing	191
10.6.	Cascade Processes and Multidimensionality	195
10.7.	Conclusions	205
10.8.	Appendix: The Fractal Sums of Pulses Process	207
Chapter 11.	Processing Satellite Infrared and Visible Imagery for Oceanographic Analyses	213
	P. E. La Violette	
11.1.	Introduction	213
11.2.	Infrared and Visible Ocean Imagery	215
11.3.	Satellite Image Analysis Techniques	222
11.4.	Examples of Processed Satellite Imagery: Alone and With Conventional Data Oceanographic Analysis	232
11.5.	Summary	238
Chapter 12.	Image Processing In Optical Astronomy	243
	J. J. Lorre	
12.1.	Introduction	243
12.2.	Calibration	244

12.3.	Enhancing Faint Surface Brightness Features	249
12.4.	Automated Target Detection & Extraction	251
12.5.	Geometric Transformations	255
12.6.	Colour Space Transformations	256
12.7.	Two Dimensional Histograms	260
12.8.	Polarisation	262
12.9.	Resolution Restoration	264
12.10.	Rotation Curve Determination	266
12.11.	Conclusion	267
Index		271

1
Computing issues in digital image processing in remote sensing

Dr Jan-Peter Muller,
Department of Photogrammetry and Surveying,
University College London,
Gower Street, London WC1E 6BT, UK
ARPANET:jpmuller@uk.ac.ucl.cs

1.1. Introduction.

Scant attention has been paid so far in the published literature to a critical analysis of the computing issues involved in the development and application of digital image processing systems in remote sensing. One of the few exceptions is the text by Green (1983) which contains sections on image cataloguing and the then currently available NASA software for image processing. This chapter attempts to provide a critical subjective analysis of what the author believes to be the key issues in this area.

Over the last thirty years, image processing has moved from being a subject area dominated by electronic engineers and computer scientists with access to powerful mainframe resources to one that includes schoolchildren playing with educational software (i.e. games) on small micro-computers.

In the civilian area, image processing was first applied by Bob Nathan and co-workers, in an operational sense, to the restoration of badly degraded analogue images returned by the first landers and orbiters to the Moon. This work in planetary science which began at the Jet Propulsion Laboratory in the early 1960s continues to the present-day with applications in the Voyager fly-by of Neptune, Magellan mission to Venus (incorporating a SAR imager), Galileo mission to Jupiter (incorporating a CCD Solid-State Imaging (SSI) and Near-Infrared Mapping Spectrometer (NIMS)). Applications of some of these techniques to astronomical images are described in Chapter 12 by Jean Lorre.

Before discussing the computing issues involved in applying these techniques it is worthwhile spending some time to examine the philosophy behind the computer processing of remotely-sensed images.

Image Processing Philosophy.

Figure 1.1 shows a schematic diagram of the classical model for image processing tasks. Image processing is usually thought of as image to image transformation operations which change grey-level values in an array pixel by pixel. However, we include image to sparse data and sparse data to image operations as they are all frequently referred to as image processing.

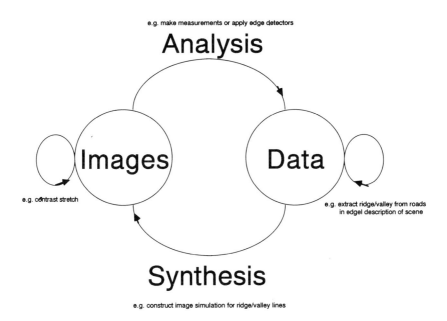

Figure 1.1. Image processing philosophy

Image to image transformation operations are performed in order to enhance the perceived information content of an image through, for example, Look-Up Table (LUT) manipulations of the image histogram (Castleman, 1979). In some cases, such as radiometric de-calibration of the camera images, these LUT operations may be used to adjust all the pixel grey-level values to some standard value of I/F (see, for example, Danielson *et al.*, 1980, Markham & Barker, 1987). In this case, the resultant grey-levels will be saved. In other cases, fine detail in areas such as self-shadows may be brought out through a temporary histogram equalisation (see last section).

The second kind of operation shown in figure 1.1 is image to sparse data or analysis which involves the extraction of some aspect of the image either manually or by the machine. The former case includes examples such as manual measurements of cloud position in a pair of images separated by some interval of time (see Muller, 1982) whilst the latter case includes automated extraction of edgel information (such as coastlines, loc.cit.).

The third kind of operation is very similar to the preceding two in that data is processed. However, in this case a further level of description is extracted by processing the data either manually or automatically. An example of the former is where a road feature is tracked in a SPOT image using the edge pixels (edgels) derived from a panchromatic image superimposed on a false colour composite of a multispectral image (see Stevens *et al.*, 1988). An example of the latter is where a connected component labelling is used to join all the edgels into some vector data structure (see, Ballard & Brown, 1982).

The final operation which comes under our generic title of image processing is the synthesis of images from sparse data historically referred to as computer graphics, but nowadays called visualisation (see Wolff, 1988). In this case, the machine must do all the processing to interpolate the sparse array. Examples include cloud simulations (see chapter by Schertzer & Lovejoy) and dynamic landscape visualisations (see Muller *et al.*, 1988d).

What's special about Digital Image Processing in Remote Sensing?

Digital image processing in remote sensing can be characterised by a number of computer processing-related constraints, including:

1. input image sizes are larger than standard CCIR or NTSC CCD images used in robot vision and automated manufacturing applications;

2. input images are frequently multispectral, but multitemporally unregistered;

3. number of operations per pixel are usually small;

4. little if any information can be extracted from remotely-sensed images without computer processing owing to the severe information loss incurred in producing false colour composite hardcopy.

There is still a number of commercial systems which is actively promoted in the remote sensing community which barely, if at all, address these problems. You can recognise them at the exhibition demonstration by the time it takes for each image to be displayed (a recent example boasting a RISC processor took almost 10 minutes to load and display an image from disk) or the time it takes you to pan, scroll, or zoom in and out (anything less than refresh rates and you should walk away) or the time it takes to calculate and display a histogram.

In addition to these computational constraints, psychophysical experiments have shown that there are a number of human factors which must be taken into account when assessing whether a particular image processing system will be practicable for remote sensing. These factors include:

1. feature perception is limited by the spatial resolution of the display device (see discussion on MTF characteristics in Pratt, 1978);

2. feature perception is also limited by the range of contrasts and black and white vision is usually limited to only being able to distinguish some 16 grey-levels (loc.cit.);

3. people are only prepared to wait a few seconds (at most) for feedback from an image processing operation involving human intervention, otherwise they will do something else (e.g. read their electronic mail, look at another image, talk to someone).

These practical ergonomic constraints are rarely, if ever, addressed by equipment manufacturers who force practitioners to sit in darkened rooms at extremely close range to multiple dark screens with poor convergence. However, there is some light at the end of the tunnel in the form of new workstation technology which has been designed to work in offices and even homes (see later section).

Let us now examine these computing issues in some more depth starting with hardware including computer requirements and display and interaction technology; software including image processing languages, libraries, packages and human-computer interface issues and in the age of WIMPs (windows, icons, menus, pull-downs) some specific examples of a recent development in image processing using workstations.

1.2. Hardware Issues.

Computer Requirements.

In the old days when image storage was very expensive (i.e. rooms were filled with 32K of RAM) and before the age of virtual memory management systems, the system engineers who developed image processing techniques devised clever software in machine code or assembler for taking an image from magnetic tape, partitioning it into many small blocks and processing each block sequentially using this tiny amount of available memory. These were the glorious days of card-oriented batch programming when the card reader swallowed and sometimes destroyed 2 metres of cards and high quality output consisted of character over-printing on lineprinter plots (for examples see Hord, 1982).

Display Processing Paradigm.

At the end of the 1970s, semiconductor memory became cheap and fast enough to build framestore systems to attach to computers. Around the same time a few computer equipment manufacturers realised that to meet the demands of cheaper RAM a higher speed input/output (I/O) channel was needed to take data from slow storage devices, such as Winchester magnetic disks via the CPU to the image displays. The fusion of the two technologies, together with virtual storage techniques to page data onto swap space on disks meant that the huge volumes of data pouring down from Earth-orbiting satellites could be processed up to a theoretical limit imposed by the addressing space of the computer (4.2GB for 32-bit processors). The age of digital image processing of remotely-sensed data had begun.

An example of this combination of virtual memory management with actively buffered framestores was the Interactive Planetary Image Processing System (IPIPS) developed at University College London in the early 1980s and later on transferred to Imperial College (see Hunt et al., 1985). Figure 1.2 is a schematic diagram of the hardware configuration.

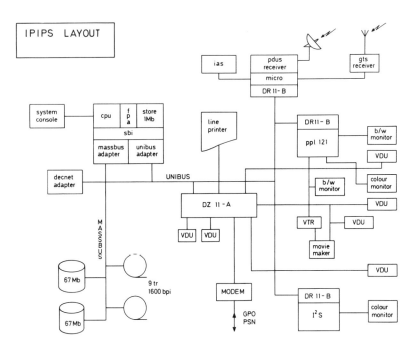

Figure 1.2. IPIPS schematic layout

Four points should be noted on this system. Firstly, massive data storage was insufficient to keep up with the 25MB of Meteosat data which poured down from that satellite every 30 minutes so that an Image Archiving System (loc.cit.) became necessary to keep up with the data-rate. Secondly, an analogue video-disk was used to act as a passive buffer for time sequences of satellite images owing to the mis-match between the I/O ($<1.3\text{MBs}^{-1}$) speed of the Unibus and Massbus and the refresh-rate of cloud motion sequences ($15 MBs^{-1}$). Thirdly, there was a physical separation between command line interaction on VDUs, image display on the I^2S and interaction devices (such as track-balls and a digitising tablet) which caused several ergonomic problems (these were mainly overcome by the on-screen interactive system described in the chapter by Tildsley and England). Finally, the system was reasonably well load balanced for images of 512×512 pixels up to four multispectral bands between processing on the minicomputer VAX (for batch jobs involving unitary transforms or geometric resampling) and processing using the simple pipeline Arithmetic Logic Units (ALUs), LUTs, zoom, pan and scroll facilities of the display processor. A comparison of speeds for simple arithmetic operations (with 10-bit precision on the I^2S and 32-bit precision on the VAX) shows that the display processor operated at around 6.5Maops (arithmetic operations per second) compared with around 100Kaops for the host machine.

This type of mixed processing environment has been very common in research facilities around the world involved in remote sensing. However, even when improvements have been made in the host processor technology, this approach suffers from a number of shortcomings:

1. the specialist display processor is usually very difficult to programme;

2. the limited arithmetic precision of the ALUs (usually <12-bits) severly restricts the number and types of operations that can be carried out without resorting to the much slower host;

3. the world does not consist of 512×512 pixels and the limitations imposed by a serial pipeline processor means that advances in multi-tasking window-based environments cannot be exploited.

Parallel Processing Paradigm.

Around the same time as the IPIPS system was being developed, Michael Duff and co-workers in the Image Processing Group at UCL were designing and building a prototype processing array consisting of $96 \times 96 \times 1$-bit processing elements (PEs) operating as a Single Instruction Multiple Data (SIMD) system (see Flynn, 1966) known as the Cellular Logic Image Processor (CLIP4).

CLIP4 (Duff, 1983) addressed several problems inherent in previous systems, viz:

1. with one pixel per PE, complete images (albeit $96 \times 96 \times 1$-bit) could be processed in sub-refresh-rates;

2. with dual-ported RAM, results could be displayed as they were being processed;

3. with nearest 4- and 8-way neighbour connectivity, the higher level of data abstraction could be addressed using bottom-up morphological analyses (such as erosion/dilation) or region-growing (loc.cit.).

However, several problems remain and new problems have emerged from a study of such systems, viz:

1. SIMD systems are difficult to programme, particularly for floating-point arithmetic operations to extract the processing speed;
2. problems need to be decomposed so that they fit onto regular small SIMD arrays using tiling algorithms;
3. the I/O speed to such arrays is still slow and currently limited to the host bus bandwidths (Sun workstation disk SMD controllers limited to 2.5MBs^{-1}) so that video-rate processing for complete frames cannot yet be accomplished.

For many applications involving repetitive processing to extract low-level features such as cells in medical slides, these type of systems are ideal in that processing to display rates are not critical. However, in remotely-sensed image processing the human analyst is still a key player in the processing loop when trying to interpret the information content of a satellite or aircraft scene.

A recent solution to many of these hardware problems has come with the development of more flexible Multiple Instruction Multiple Data (MIMD) systems (Flynn, 1966) based on 32-bit microprocessor arrays of Inmos transputers (see Muller *et al.*, 1988a). The key breakthrough with these devices has been the simple glue logic consisting of built-in bidirectional asynchronous serial links operating at 20Mbps. Floating-point operations are up to ten times faster than a VAX minicomputer and arrays can be built with up to around 1K transputers using high-speed silicon switching technology (loc.cit.).

Figure 1.3 is a schematic diagram of a 22 T414 (integer transputer) processor array, which has been used for research into image processing at RSRE Malvern and UCL. Although Muller *et al.* (1988) have shown that this processor array is not very fast for low-level edge detection operations (see Table 1, loc.cit.) its performance on CPU-intensive operations such as Fast Fourier Transforms (Roberts, 1988) is comparable to SIMD machines with more than 32 times the number of PEs.

For certain classes of image processing operations such as stereo matching (Muller *et al.*, 1988a) where upwards of 1000Maops (Muller *et al.*, 1988b) may be required to process data at video-rates, the transputer array appears to offer the only current realistic solution. The processing speed is currently not matched by the I/O speed (around 6MBs^{-1}) for the machine shown, but recent developments in multiplexing up to 64 transputer links will give potential I/O bandwidths of up to 200Mbps (video-rate for 13-bit data). The faster GaAs 1um technology should come available in the future should provide up to two orders of magnitude increase in speed per processing element so that reasonably small arrays can operate at refresh-rates for very CPU-intensive tasks. Finally, special-purpose graphics accelerators with interfaces to processor arrays running at up to 250Mbps (e.g. Sun

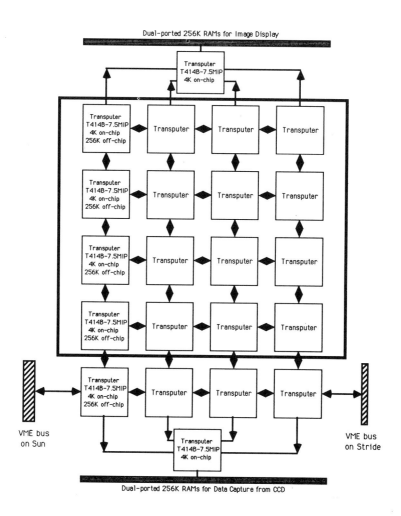

Figure 1.3. RSRE Meiko

TACC-1, see England, 1988) may allow the real-time display of results from such image processing engines on a standard workstation.

CD-ROMs, WORMs and the data mountain.

Only a very small percentage of the total data collected from Earth observation satellites has ever been examined let alone digitally processed. One of the reasons for the development of the so-called data mountain is the lack of a coherent processing strategy which would allow us to retrieve information in a format which could be used directly in operational forecasting models. Exceptions are wind-fields derived from geostationary satellite images using cross-correlation and wheat acreage estimates derived using LANDSAT and a combination of other data sources. In almost all other areas of economic activity including thematic mapping, data is enhanced and then usually written to film and filed somewhere.

If we can assume that processing algorithms are sufficiently robust and the satellite sensors well enough designed to meet our application goals, then advances in parallel processing may in future permit us to process the data at the same rate as we receive it from the satellite. In the meantime, the shift towards a holistic view of planet Earth (Gaia) planned for the late 1990s with the NASA, ESA and Japanese polar platforms with multi-disciplinary teams of scientists processing global data-sets means that we will be forced to think about how we might process data online and how we might distribute it to all the hundreds of scientists and value-added industries which will emerge.

If this future scenario were based on magnetic tapes, it would probably be more sensible for them not to launch the platforms (ESA do not appear to understand this problem as ERS-1 distribution is to be tape-based). The problems with tapes can be stated as follows:

1. they leak data from the time they are written, unless they are mounted, spun and kept in air-conditioned warehouses;

2. they are extremely time-consuming to load, catalogue and write;

3. they are slow and unreliable and frequently tapes written on one computer system cannot be read by the same type of computer at another installation;

4. they are messy to handle, particularly if the tape runs off the spool;

5. hardware faults on tape-drives sometimes cause total loss of data.

Fortunately, a solution has appeared on the horizon in the form of new optical storage media (see chapter by Ferns and Press) based either on Write Once Read Many (WORM) times 12" laser-disks (storage up to 1.2GB per side) which can be used for archiving and backup or for distribution of data by the satellite agencies, CD-ROM based on 4.72" Compact Disks (storage up to 550MB). The convenience of such systems based on juke-boxes has been recognised by NASA in its plans for online storage of all the data contained in its planetary archive but has yet to be recognised by any other group.

Display resolution and interaction devices.

Many existing image processing systems are based on the results of psychophysical experiments in the 1960s which indicated that about 7 bits were needed for monochrome pictures. It is often assumed by inference (particularly by image processing equipment manufacturers) that therefore three times as many bits are required for colour.

Cowlishaw (1985) gave very strong evidence to show that neither of these two conclusions are true for a large variety of different imagery. In particular, he showed that only 4-bits (16 grey-levels) were needed for monochrome picture display and 8-bits (256 colours) for colour display (4-bits Green, 2-bits Red, 2-bits Blue). Furthermore, he showed that this could be achieved using simple error diffusion algorithms (see review in Ulichney, 1988). This means that cheap 8-bit display devices used for personal computers with less than 4 pixels/mm display resolution, such as the IBM Professional Graphics Adapter (see Myers & Bernstein, 1985) and the Sun workstation (see Muller *et al.*, 1988c) can be used to display colour images and up to two 4-bit grey-scale images (see discussion in last section).

Figure 1.4 shows the effect of this error diffusion algorithm (based on Floyd and Steinberg, 1976) on the display of a panchromatic SPOT image. Examination of output such as this figure shows that cheap laser printers (with 12 pixels/mm resolution) can now be used as hardcopy devices for remotely-sensed images using these error diffusion algorithms.

Manual interaction with an image, such as the measurement of position or variation of LUTs can be accomplished using a variety of different interaction devices. These devices for pointing at features on image display devices have hardly changed since the late 1960s when joysticks became commonplace on vector storage scopes for signal analysis. These devices can be categorised as coarse-scale pointing (such as single button mice on personal graphics computers such as the Apple Macintosh or touch-screen interfaces), medium-scale pointing (such as trackball cursors on many commercially available image processing systems and light-pens on previous generations of image display equipment) and fine-scale pointing (such as the glass-fronted pucks used for cartographic digitisation).

Little quantitative assessment of the spatial resolution limits of these different devices with different display resolutions appears to have taken place to date. This seems to be symptomatic of the lack of interest that computer peripheral manufacturers have in human-computer interfaces (HCIs). However, experiments conducted at UCL by Dr V Paramananda have indicated that even with pixel-replicated zoom, consistent measurements of well-defined points can only be repeated to an accuracy of around 0.3 pixels r.m.s. using a 3-button mouse on a Sun-3 colour workstation.

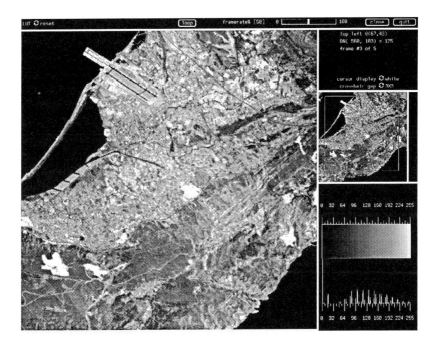

Figure 1.4. Error diffusion (disp display)

1.3. Software Issues.

Languages, Libraries and Packages.

The variety of different techniques available in image processing (see, for example the range described with specific reference to remote sensing in Moik, 1980) has led to the blossoming of a huge variety of incompatible software with different user interfaces, different data formats, different history functions and different programming languages. Preston (1983) provides an excellent review of some 42 image processing languages in use before 1983. In addition to the languages discussed in this list, there are only a few significant ones which have achieved widespread use, mainly associated with particular display processors or firmly rooted in a particular operating system ethos (see description of HIPS later on).

To qualify for being called a language, the routines must have a control structure which allows a syntax to be defined in a similar way to a compiled programming language but with data structures which are image arrays. Examples described in this book include TAE (see chapter by Perkins *et*

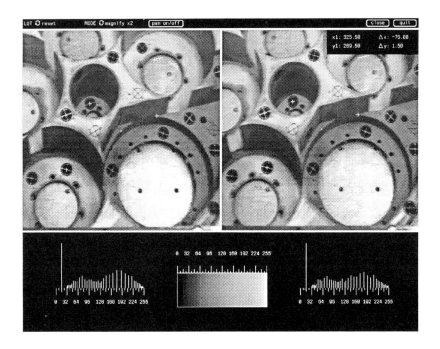

Figure 1.5. Stereo display

al.), IAX (see chapter by Jackson) and VICAR (see chapter by Lorre).

Two approaches have been taken to the development of image processing languages. In the first, a subroutine library of portable routines is defined with a clean interface. SPIDER (Tamura *et al.*, 1983) represents one of the most comprehensive of these classes. In the second approach, a software package with a full range of image data-base functions (see definition in Green, 1983), a high-level command language and a powerful range of image processing functions is available sometimes for a variety of different data types. A recent example of this class is HIPS (Landy *et al.*, 1984) which is built on the piping methodology of Unix where output from one function can become the input to the next function in the list. HIPS can handle a variety of different data types (byte, halfword, floats, complex) and data structures (e.g. time sequences of multispectral stereo) which very few byte-oriented languages can because it is written in the 'C' programming language wherein memory can be dynamically allocated (using pointers) and new data structures can be easily constructed. Limitations of this approach are currently concerned with image processing operations which may produce more than two outputs or need more than two inputs although the use of intermediate files can be made.

One of the few languages which has recently had a re-birth is VICAR (for description see Castelman, 1979) because of its extremely frugal use of memory. Versions of this language were first ported from the IBM mainframe environment in the early 1980s to a VAX/VMS system by Hunt *et al.*, 1985. More recently, Jean Lorre and Joel Mosher of Pasadena (California) have ported VICAR to a Unix operating system (where the command syntax and grammar is an extension of the command set available in the OS) and over the last two years, VICAR has been modified so that it can run on any of the IBM PC (or XT or AT) and PC-clone machines running DOS3.3 or higher†.

Human-Computer Interfaces.

Computer users, ever since the introduction of Visual Display Units (VDUs) have developed Human-Computer Interfaces (HCIs) which attempted to mimic some aspect of human dialogue. The traditional view holds that as the machine is a dumb servant of a human user, terse command-level dialogue should be the most universally acceptable HCI. However, as computers have penetrated a wider audience, it is becoming clear that many users have neither the time nor the inclination to learn or try to understand the logic behind most command-level interfaces. In addition, it appeared that repetitive data entry tasks were better served using a fixed form menu rather than a succession of command-question/answers. Latterly, as both the syntax and the level of functionality have become more complex (so increasing the alienation of the computer from ordinary users) attempts have been made to address the novice user (for novice, read occasional) through the devlopment of tutor and menu procedures to wean the novice user into the design ethos of the chosen system (see discussion of TAE in Chapter 3).

In the late 1970s at the Xerox Palo Alto Research Center (Smith *et al.*, 1982) a new style of HCI was developed which used the metaphor of pictographic symbols or icons to represent everyday objects such as sheets of paper, folders and trash cans (waste paper baskets). Gittins, 1986 reports that very few systematic studies (only one is quoted) have been done to try to compare icon, menu and command interfaces to the same computer system (one hopes that this situation will dramatically improve soon considering its importance). Meanwhile, the number of WIMP interfaces has burgeoned after the introduction by Apple in the mid-1980s of their Macintosh microcomputer. This has been followed by WIMPs on DOS-based microcomputers (such as Microsoft Windows) and most recently, by Look

---

† A free machine-readable source copy of this software together with an online index is available for research workers involved in non-commercial work with access to the ARPANET, JANET, BITNET or EARN networks at ARPANET email address pcvicar-request@cs.ucl.ac.uk. No correspondence will be entered into unless emailed to this address.

and Feel on Unix systems.

In all of these cases, these HCIs have one theme in common, they appear to be easy to use for novice users and are in the author's limited experience frequently resented and derided by more experienced users. In the last few years, WIMPs have become increasingly synonymous with multitasking environments where the windows are used to represent different processes either on the same machine or more commonly on different machines on the same local or wide area network. Stern, 1987 reviews the main characteristics of the fixed Macintosh-based system and compares them with the two competing multi-machine WIMP standards, X11 (developed at MIT with funding from DEC) and NeWS (developed at Sun Microsystems).

The relevance of these window standards to digital image processing is based on the author's perception that graphics workstations are now becoming so widespread and inexpensive that serious attention needs to be paid to their possible use as image display terminals. Many of these workstations today can be used to perform almost all of the tasks previously considered the exclusive domain of display processors (see next section). In this respect, the NeWS windowing standard appears to be a more flexible system for image processing as its imaging model is not based on any particular pixel resolution.

The advent of the integration of multifarious workstations and personal computers in heterogeneous computing environments such as the one shown in figure 1.6 suggests that hardware manufacturers will need to consider in much more detail the needs of a user community which cannot understand why brand X of a particular computer company, Y cannot communicate with brand A of another company, B. The transparent transfer of images (no matter what the bit ordering) from one machine to another will continue to be far away from reality until windowing standards emerge.

1.4. Image Processing Workstations—an Examplar.

Image display.

The first assumption that needs to be made about future image processing hardware is based upon the author's firm belief that a single unified viewport into image, graphics and fonts is vital to the success of integrating remotely sensed data with existing spatial information databases. The preponderance of different screens is not only poor ergonomics but also means that image processing is less likely to be as commonplace as many people hope. The key to this is to take the image display out of the closet and into everyone's office so that processing pictures becomes as natural as reading one's electronic mail.

The difficulty is that although engineering graphics workstations address

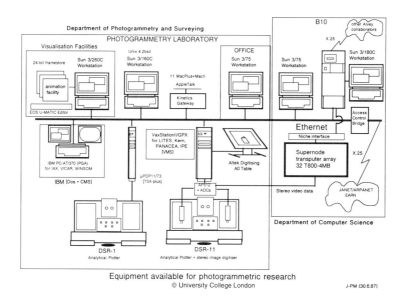

Figure 1.6. UCL IP equipment

many of the budgetary, ergonomic and technical problems, the lack of standardisation between equipment manufacturers means that not everyone will have access to the same resources and that different workstations will suit different types of uses. At a more prosaic level, different manufacturers have conspired to build different bit representations or even logical-to-physical magnetic tape block mappings. The continued use of magnetic tapes perpetuates many of these problems unnecessarily as mass storage, such as CD-ROMs, have already put draft standards for data interchange onto the ISO committees.

Even when workstations can connect to the same physical medium, such as co-axial Ethernet, this does not mean that they can communicate with one another if they have incompatible operating systems (see figure 1.6). There are many ad hoc solutions available from companies for throwing ropes across the operating systems chasm, but frequently theses ropes break at the slightest pressure.

Once a solution is found, however, to these technical hardware problems, the benefits of a workstation environment over a single dedicated display processor on a time-shared minicomputer are enormous. Data can be taken from a wide variety of different input sources, including digitised maps and photographs, field instruments, data networks and mass storage devices and

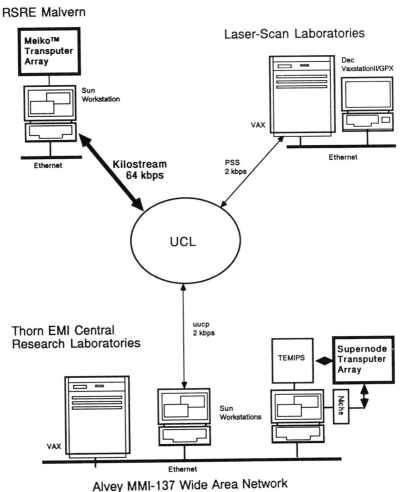

Figure 1.7. Alvey WAN

be immediately available to the whole research group.

In addition to this open access to a distributed filestore using vendor solutions such as Sun-NFS, processing may also be executed remotely on one or more machines, using a vendor solution such as Sun-RPC. This means that a large classroom of workstations could transform out of class hours in to a distributed supercomputer and the single CPU-bottleneck could be overcome. This has been used at UCL to reduce stereo matching computations down from 40 CPU days to a few hours (see Muller *et al.*, 1988b).

The WIMP-HCI has been coupled at UCL using a "look and feel" based on the notion of a fixed canvas with fixed function. One of the great conceptual and design breakthroughs of these types of interfaces in the early 1980s was the realisation that once you had convinced a user that they liked using a system, you made sure that the next application software the user tried had many of the same commands. In this way, confidence was built up by the user in their abilities and the uniformity of interface provided a powerful constraint on programmers.

Almost all image processing workstations require the same type of basic functionality. This includes flicker, pan, scroll, zoom-in and zoom-out and Look-UP Table (LUT) manipulation including histogram equalisation, single point transfers, pseudo-colouring. Display processor manufacturers use special purpose chips to perform these tasks at video refresh rates. They dont need to because almost all of them can be done with any colour workstation using the existing bit-mapped technology.

Automated histogram display for whole images and regions-of-interest is another important aspect as well as feedback for the effect of any LUT operation, not via a transfer curve but directly on a grey-scale wedge on the screen. If you examine figure 1.4, you will see all of these aforementioned features together with sub-pixel positional information on the location and local grey-value of the pixel under the cursor. In addition to pull-down menus, the programme can be operated by an experienced user (read hacker) using single keystrokes commands.

Image-image Registration and Stereo Measurement.

The next logical step from a single image display is to attempt to register two images together or find the common tiepoints between them. In the past this has involved split-screen displays that have occupied many unhappy hours for many a user. With bit-mapped displays, two grey-scale images can now be combined using logical OR operations on error-diffused 4-bit data. The advantages are obvious if the tiepoints are differently coloured on the two images as features can be scrolled and zoomed-in or out over each other. The visual impact and subsequent measurement accuracy is much higher.

The final extension of this image-image registration paradigm is shown in figure 1.5 for stereo image data. In this case, a mirror stereoscope is currently required to help the human visual system to fuse the two images. However, the common features of the interfaces and ease of use means that this program has been used for a far wider variety of applications than originally intended.

1.5. Conclusions.

An example of a multifarious workstation environment of the present-day is shown in Figure 1.6. All of the systems are able to talk to each other to varying degrees of sophistication. The use of windowing standards will help to unify the user interfaces across the system so that the "look and feel" of any remote sensing programme will be as uniform as possible.

Looking towards the future, the advent of group supercomputer resources with arrays of transputers and the increased internetworking between sites means that the workstation/distributed processing paradigm is likely to spread into Wide Area Networks (see figure 1.7). The obvious benefits to increased computer power at much lower cost means that the image processing in remote sensing will be helped to spread beyond the current confines of the well-established laboratory and into all sectors of education, research and development.

References.

Castleman, K.R., 1979 *Digital Image Processing*, Prentice-Hall, New Jersey.

Danielson, G.E., Kupferman, P.N., Johnson, T.V., Soderblom, L.A., 1981, Radiometric performance of the Voyager cameras, *J. Geophys. Res.*, vol.86(A10), 8683–8689.

Duff, M.J.B., 1983, *Computing structures for image processing*, Edited by M.J.B. Duff, Academic Press, London.

England, N., 1986, Application accelerator: development of the TACC-1, *Sun Technology*, Winter 1988, 34–41.

Floyd, R.W., Steinberg, L., 1976, An adaptive algorithm for spatial greyscale, *Proc. Soc. Inf. Display*, vol.17(2), 75–77.

Flynn, M.J., 1966, Very high-speed computing systems, *Proc. IEEE*, vol.54, 1901–1909.

Gittins, D., 1986, Icon-based human-computer interaction, *Int. J. Man-Machine Studies*, vol.24, 519–543.

Green, W.B., 1983, *Digital Image Processing, A Systems Approach*, Van Nostrand Rheinhold, Wokingham.

Hord, R.M., 1982, *Digital Image Processing of Remotely Sensed Data*, Academic Press, New York.

Hunt, G.E., Barrey, R.F.T., Clark, D.R., Easterbrook, M., Gorley, R., Marriage, N., Muller, J.-P. A. L., Roff, C.E., Rumball, D., 1985, The Interactive Planetary Image-Processing System, *IEEE Trans. geosci. Rem. Sens.*, vol.GE-23(4), 581–595.

Landy, M.S., Cohen, Y., Sperling, G., 1984, "HIPS, A Unix-based image processing system" *Computer Vision, Graphics and Image Processing*, vol.25, pp.331–347.

Markham, B.L., Barker, J.L., 1987, Thematic Mapper bandpass solar exo-atmospheric irradiances, *Int. J. Rem. Sens.*, vol.(3), 517–525.

Morris, A.C., Stevens, A., Muller, J-P, 1988, Ground Control determination for registration of satellite imagery using Digital Map data, *Photogrammetric Record* (in press).

Muller, J-P A L, 1982, *Studies of Jovian Meteorology using Earth-based and spacecraft data*. Ph.D. Thesis (unpublished) University of London.

Muller, J-P., Collins, K.A., Otto, G.P., Roberts J.B.G., 1988a, Stereo matching using transputer arrays, Invited Paper, ISPRS 16th Congress, Kyoto, Japan, July 1–12, 1988. *Int. Arch. Photo. Rem. Sens.*, vol.27–A3, 28pp.

Muller, J-P., Otto, G.P., Chau, T.K.W., Collins, K.A., Dalton, N., Day, T., Dowman, I.J., Jackson, M.J., O'Neill, M.A., Paramananda, V., Roberts, J.B.G., Stevens, A., Upton, M., 1988, Real-time stereo matching SPOT using transputer arrays, *ESA SP-284*, IGARSS'88, Edinburgh, 13–16 September 1988.

Muller, J-P., Paramananda, V., Richards, S., 1988c, Graphics workstations for integrated data collection, simulation and display, Contributed paper ISPRS 16th Congress, Kyoto, Japan, July 1–12, 1988. *Int. Arch. Photo. Rem. Sens.*, vol.27–A3 (abstract only).

Muller, J-P., Day, T., Kolbusz, J., Dalton, M., Richards, S., Pearson, J.C., 1988d, Visualisation of topographic data using video animation, Contributed Paper, 16th ISPRS Congress, Kyoto, Japan. *Int. Arch. Photo. Rem. Sens.*, vol.27–A3, 12pp (also in Chapter 2).

Myers, H.J., Bernstein, R., 1985, Image processing on the IBM Personal Computer, *Proc. IEEE*, vol.73(6), 1064–1070.

Pratt, W.K., 1978, *Digital Image Processing*, John Wiley & Sons, New York.

Preston, K. Jr., 1983, "Progress in Image Processing Languages" Chapter 13 in *Computing structures for image processing* Edited by M.J.B. Duff Academic Press.

Roberts, J.B.G., 1988, Private communication.

Smith, D.C., Irby, C., Kimball, R., Verplank, B., 1982, Designing the Star user interface. *Byte*, vol.7, 242–252.

Stern, H.L., 1987, Comparison of Window Systems, *Byte*, vol.12(13), 265–272.

Ulichney, R.A., 1988, Dithering with blue noise, *Proc. IEEE*, vol.76(1), 56–79.

Tamura, H., Sakane, S., Tomita, F., Yokoya, N., Kaneko, M., Sakaue, K., 1983, Design and Implementation of SPIDER—A transportable image processing software package, *Computer Vision, Graphics and Image Processing*, vol.23, pp.273–294.

Wolff, R.S., 1988, Visualisation in the eye of the scientist, *Computers in Physics*, vol.2(3), 28–35.

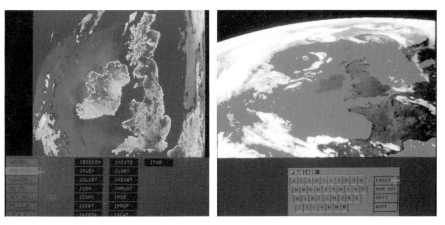

Figures 4.1 (left) and 4.2. (right). Figure 4.1. Top level menu. The figure illustrates a top level menu. Any of the programs may be chosen to give a further menu displaying the program's options. Alternatively, **HELP** *may be chosen to give simple details on the programs available. The image shown is a NOAA 8 image coloured to show temperature difference across Britain. (Red is hottest). Figure 4.2. Alphanumeric input. As well as a file name being selected from a listing of file names, the file name may be entered directly using the menu keyboard. The IMI also polls the terminal keyboard, enabling real keyboard input if desired. The image is a frame from a movie of processed METEOSAT images, as supplied daily to Thames News at Six, ITN News at One, and BBC Breakfast Time.*

Figure 7.1. LANDSAT colour composite

Figure 9.2. LANDSAT mosaic (MSS4, MSS5 and MSS6)

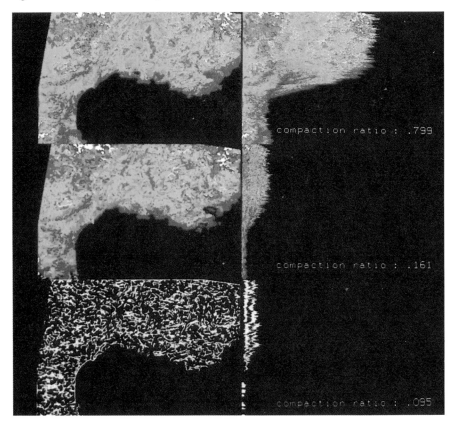

Figure 9.3. The Run Length Coding of different data.

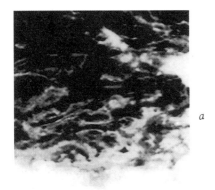

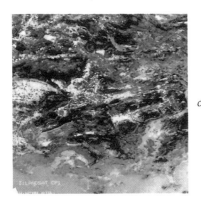

Figure 9.4abc. The use of LANDSAT/HCMM data: (a) HCMM night temperature data registered to LANDSAT with 80m resolution (b) LANDSAT false colour of MSS6, MSS5 and MSS4 with 80m resolution (c) IHS display of LANDSAT/HCMM data, LANDSAT 1st principal component as Intensity, HCMM night temperature as Hue, Saturation constant.

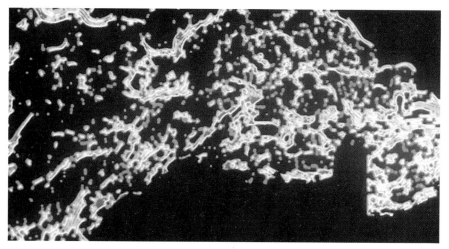

Figure 9.7b. The detected high-gradient area on the gravimetric data.

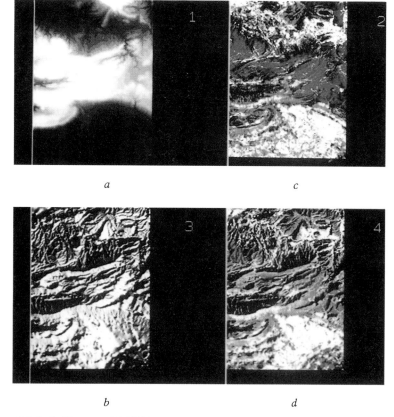

Figure 9.8abcd. The use of LANDSAT and DTM data: (a) Image of the Digital Elevation Model (b) Image of the synthetic reflectance (c) LANDSAT data in false colour (d) Modulated LANDSAT data in false colour.

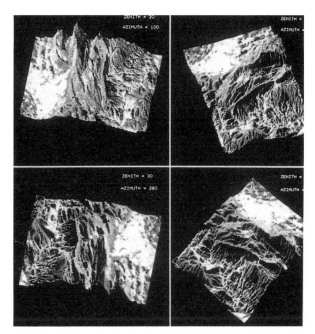

Figure 9.8ef. (e) Three-Dimensional views of LANDSAT data (f) Pair of stereo views of LANDSAT data

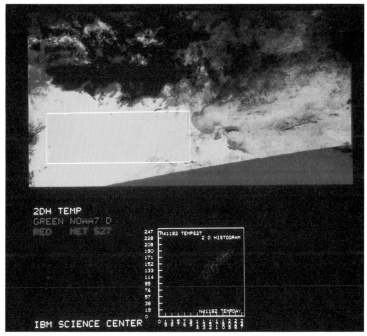

9.9. NOAA AVHRR ground temperature image and METEOSAT top-of-the-atmosphere temperature image over Tunisia (top). The two dimensional histogram of these two temperatures over the the zone defined by the rectangle (bottom). A highly linear correlation is evident.

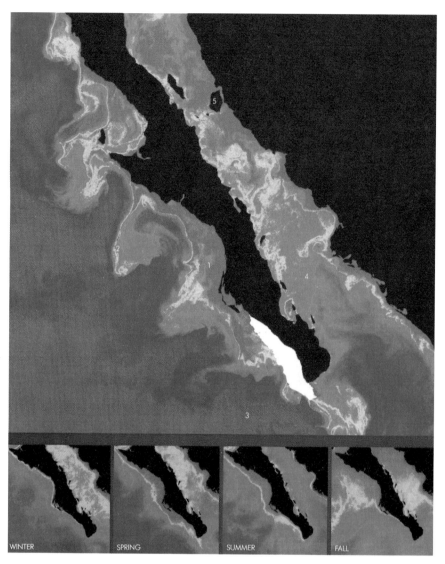

Figure 11.3. The image shows ocean colour data collected by Nimbus-7 CZCS at near local noon on May 12, 1979. Although the values show the chlorophyll distribution in the ocean, they are equally instructive in showing a region's current and frontal distribution. The four smaller CZCS images illustrate the region's seasonal variations in phytoplankton distribution. These data were collected at the NASA Goddard Space Flight Center; processing and analysis were performed by Dennis Clark at NOAA using software developed at the University of Miami.

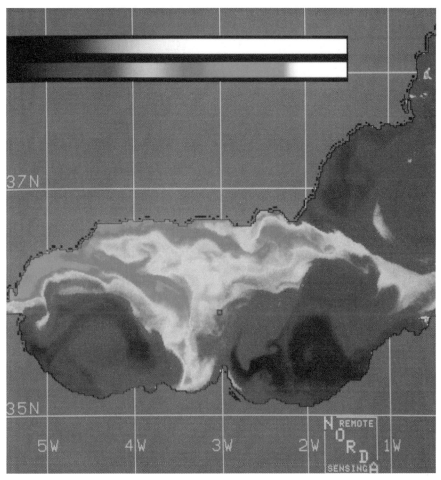

Figure 11.5. The same data as figure 11.4a with a colour assignment table to emphasize low thermal gradient ocean events.

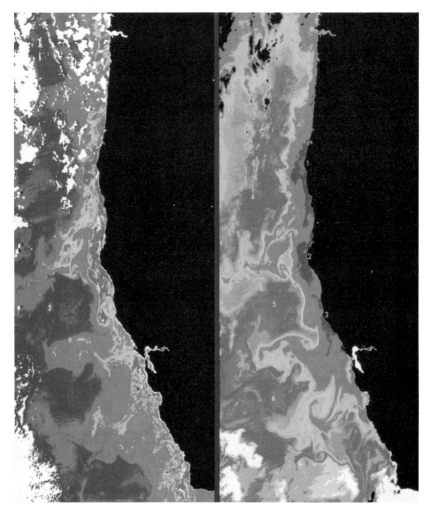

Figure 11.9. (a) A NOAA-6 AVHRR-IR image taken at approximatly 0300 hrs on July 8, 1981, showing sea-surface radiant temperatures and (b) a Nimbus-7 CZCS image of the same area taken about eight hours later showing phytoplankton chlorophyll pigment concentration. Data were received at the Satellite Oceanography Facility at the Scripps Institution of Oceanography and were processed and analyzed at the Jet Propulsion Laboratory by Mark Abbott and Phil Zion, using computer processing routines developed at the University of Miami. In the image, prevailing winds from the northwest drive coastal surface waters offshore and induce an upwelling of cooler subsurface water. In the temperature image (right) the cool, freshly upwelled water along the coast is shown in violet-purples (8–9 degrees C), especially noticeable at Cape Blanco (1), Cape Mendocino (2), and Point Arena (3). Intermediate temperature water is indicated by blues and greens, and the warmer California Current water offshore is depicted in yellows and reds (14–15 degrees C). Several large meanders of the California Current are visible (4,5,6), as well as long filaments of cool upwelled water (7), which extend 100 to 300 km offshore. The chlorophyll image (left) shows high levels of phytoplankton along the coast in reds (chlorophyll concentration greater than 10 mg/m), intermediate levels are shown in yellows and greens, and the lowest levels offshore (less than 0.1 mg/m) are purples.

2
Visualisation of topographic data using video animation*

Jan-Peter Muller, Tim Day, John Kolbusz,
Mike Dalton, Sam Richards and James Pearson.
Department of Photogrammetry and Surveying
University College London
Gower Street
London WC1E 6BT, U.K.
ARPANET/BITNET: jpmuller@cs.ucl.ac.uk

A wide range of software tools have been developed for visualising three-dimensional range data of topographic surfaces. Digital Elevation Models (DEMs), derived from photogrammetric measurements and from digitised contour maps are here used for illustration. To increase our understanding of the information content in DEMs in isolation or in combination with remotely-sensed images and to facilitate the quality assesment of automated terrain extraction techniques a number of tools have been developed for displaying perspective views of time sequences of landscapes with a variety of different surface shading models.

We consider a variety of shading models from the texture mapping of a satellite image over a DEM to attempts to represent different surface reflectances. The texture mapping problem integrates both single band (monochromatic) and multispectral (false colour of original or colour-space transformed) satellite images. Artificial shading models from simple Lambertian shading to the ray-tracing of terrains, using Bi-directional Reflectance Distribution Functions are presented.

---

* This Chapter is a revised version of a contributed paper to the ISPRS 16th Congress, Kyoto, Japan, July 1–12 1988, *Int. Arch. Photo. Rem. Sens.*, Vol.27–A3, 12pp

2.1. Introduction.

Perspective views of the surfaces of three-dimensional data are a significant aid to an appreciation of the spatial nature of natural phenomena. Examples in remote sensing include physical interpretation of satellite image brightness values; understanding of geological structures and analysis of hydrological networks and catchment areas. Examples in aerial-based topographic mapping include understanding the detailed nature of afforestation and planning issues involved in building and highway construction.

The paper map has for centuries been the primary means of visual communication of spatial data, and its properties, advantages and drawbacks have been debated for almost as long. However, apart from flight simulator projects (see, for example, Schachter, 1983, Zyda, 1988) and military route planning (Garvey, 1987), little research into the potential of dynamic perspective views of landscape has been made. It is one of the primary objectives of our research programmes to explore the issues involved in the visualisation of a wide variety of scientific and engineering data including photogrammetric measurements of industrial parts in CAD systems (Muller, 1987b) and of anthropometric data (Kolbusz, 1988).

Visualisation of scientific data has received a great deal of attention recently as a result of a US National Science Foundation study (see Mc-Cormick, 1987, Winkler, 1987, Frenkel, 1988) which presented a detailed list of scientific, technical and socio-political arguments in support of its crucial role in interpreting the results of numerical simulation experiments; understanding the massive amounts of digital data now being generated in scientific experiments and in visual communication of ideas and products to wider audiences.

In this Chapter, however, we focus our attention on the visualisation of small-scale Digital Elevation Models (DEMs) using different surface shading techniques. It should be noted that the previewing and video animation techniques described here are quite general enough to encompass most problem domains, the main difference being the representation of surface geometry in the different application areas.

Static perspective views of landscape from aircraft (usually refered to as oblique photography, see Brew, 1980) have been used extensively since the Second World War for both photo-interpretation and reconnaisance mapping. However, apart from Gelberg (1987) and some unpublished work by Quam (1987) and the Animation Group at the Jet Propulsion Laboratory Wolff (1987) little if any use has been made of perspective viewing of terrain.

The breakthrough in the spread of visualisation techniques is coming about because of two technological developments: graphics workstations and special-purpose graphics rendering hardware, such as VLSI matrix manipulation chips (so-called "geometry engines", see Zyda, 1988). These enable visualisations to be accomplished for wireframe representations of small DEMs (without hidden line removal) at video refresh rates and for simple shaded representations of DEMs at rates of approximately 1 frame per minute. Increasing RAM memories (up to 256MB) and more sophisticated "geometry" and "rendering engines" suggest that within a few years, the types of visualisations presented here will be done at video refresh-rates and we suggest will be commonplace on engineering workstations.

Three dimensional data is considered here in terms of its bounding surface rather than its existence as a volume. To produce visual representations of such bounding surfaces, we consider this as a problem of rendering (i.e. simulating the interaction of light with a 3D surface). A number of techniques has been developed in computer graphics over the last 20 years (see, for example, Rogers, 1985) for the rendering of 3D data and the most relevant of these is discussed in Section 2.4. Section 2.2 gives a definition of the objectives of our visualisation work whilst Section 2.3 discusses the various approaches which have been adopted for the mathematical representations of DEMs. Section 2.5 describes our techniques for image display and previewing and how we define our view-path for our animations as well as what equipment we currently employ. Finally, Section 2.6 suggests ways in which the work can be extended in the future. All figures shown in this Chapter are produced using the SPOT-PEPS data-set described in more detail in Day (1988a).

2.2. Objectives.

In recent years, there has been an explosion in the amount of data collected about the Earth's surface from satellite imagery. However, work on analysing this imagery has tended to concentrate on two-dimensional image processing of its multispectral content, ignoring the increasing amounts of structural (e.g. cultural) and terrain-related information which have come from narrower spectral resolution and/or higher spatial resolution.

Several authors (Woodham, 1985; Muller, 1988c; Muller, 1988d) have recently suggested an alternative approach using a computational vision model of remote sensing. At the centre of this new approach is the attempt to model image formation through computer graphical techniques to try to understand how the patterns visible in satellite images originate and hence devise sensible data processing strategies for their future automated analysis. These studies have included the automated extraction of DEMs from SPOT satellite images (see Day, 1988a and Muller, 1988a) which form one key input to image formation models, particularly in areas for which the appropriate scale of map contour data is unavailable and the cost of digitisation may be unacceptable (see Dowman, 1986).

To test our image formation models (for representing surface reflectance and/or sky radiance) as well as to assess the geomorphological veracity (see, Muller, 1988e) of our automatically-derived DEMs, we have a number of alternative strategies which can be grouped as follows:

render a synthetic satellite image(s) from the same viewpoint as the satellite with a simple pinhole camera;

render a synthetic satellite image(s) from the same viewpoint as the satellite using an exact representation of the satellite optical system and sensor characteristics (see O'Neill, 1988);

render a synthetic satellite image(s) from a perspective viewpoint;

render the actual satellite image(s) from a perspective viewpoint;

repeat items listed above using false colour representations.

2.3. Surface representation.

The input to a DEM can come from a variety of different sources. To date we have investigated the following (see examples in figure 2.1):

a) Extract from SPOT image, for comparison;

b) Interpolated manually digitised contours;

c) Manual photogrammetric measurements taken from aerial photographs;

d) Automatically stereo-matched SPOT images interpolated to the same grid intervals.

Both the digitised contours and the stereo matched data are in the form of a collection of spot elevations which do not have a data-structure amenable to easy handling or rapid rendering. Delauney triangulation (McCullagh, 1980) is therefore employed to construct a Triangulated Irregular Network (TIN) (using Laser-Scan Laboratories' MATRIX software) from which either irregular triangle data-structures can be extracted or a regular grid can be interpolated.

To build a regular triangular data-structure from a gridded DEM we divide each square (or rectangular) DEM cell into two triangles. Instead of splitting each cell in the same direction and introducing directionality, we choose at random between the two possibilities. (See example in figure 2.2)

For DEM data, this surface representation scheme is likely to remain the most efficient until suitable alternatives can be found for storing compressed versions of elevations related to the appropriate sampling densities for the terrain type (see discussion in Muller, 1988e).

There have been a number of suggestions for different sampling schemes for smooth surfaces derived from range data (see Bhanu, 1987) which are more suitable for computer manipulation. Experiments with surface fitting methods using Bernstein-Bezier curved surface patches (see Kolbusz, 1988) compacts much detail to a given surface tolerance. Our use in this work is restricted to the interpolation of DEM data. This is useful for constructing

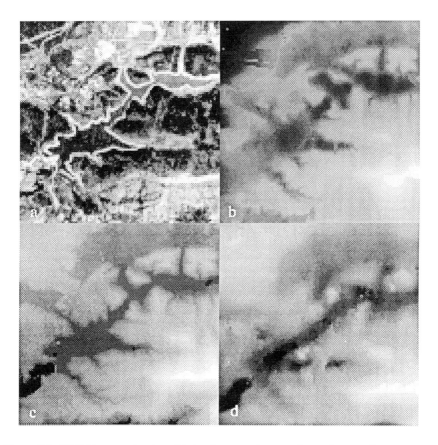

Figure 2.1. SPOT image extract, and intensity range images of area from various sources

high 'resolution' versions of DEMs from lower resolution ones to minimise spatial undersampling (the so-called 'aliasing' problem).

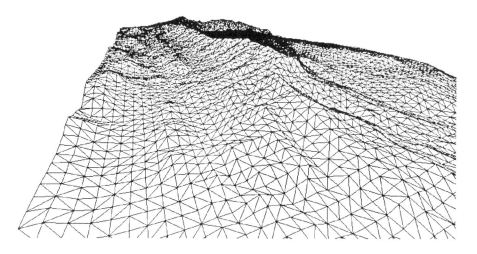

Figure 2.2. Wireframe image of low resolution DEM

2.4. Rendering.

Rendering software takes a surface representation model structure and produces images, taking into account the absolute position and relative orientation of the objects, light sources and the viewer.

Rendering software must solve a number of problems with DEM data including the removal of hidden surfaces (see review in Sutherland, 1974) and the approximate simulation of light interactions with matter (see review in Magnenat-Thalmann, 1987). These computationally expensive tasks are usually best performed on graphics workstations with specialized peripherals (see England, 1988).

Increasing levels of complexity can be used to approximate light interaction with matter (so-called shading models). Together with this complexity comes greater realism.

At the simplest level, we can assume that every facet of the surface has a constant colour. This approximation can be refined by taking into account some measure of how much the reflected light from a surface varies as a function of illumination and viewing geometry.

These simple shading models include Lambertian (Woodham, 1985) and a number of analytical approximations to the relative proportions of diffuse and specular components of scattering which have been named after their inventors: Gourard (1971), Phong (1975), Blinn (1982) and Cook-Torrance (1981).

To model light interaction between surfaces and with matter or gases in

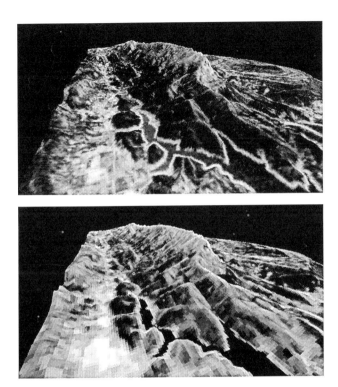

Figure 2.3. 30m DEM overlayed with SPOT panchromatic (top) and LANDSAT-TM (bottom)

the atmosphere, more advanced models such as ray-tracing, radiosity and Monte-Carlo ray-tracing must be considered (see references in Magnenat-Thalmann, 1987). These allow us to consider shadows, reflections, distributed light sources, colour bleeding and the dynamic sampling of non-stationary objects.

In the following sections, rendering is treated at two different levels of complexity: fast rendering of satellite images using texture mapping and DEMs using simple Lambertian shading and very much slower rendering of DEMs using ray-tracing to incorporate diffuse and specular surface reflectance models and sky radiance.

Fast rendering of satellite images using texture mapping.

Single-band and false colour composites of planimetrically-geocoded satellite images can be used as input to the rendering of DEMs where the grid-points or Delauney triangles are used as vertices.

Each triangle can be assigned a particular colour (at its centroid) derived from the image. Since each triangle is represented by a single colour, we are effectively reducing the image resolution down to the DEM resolution. This can be overcome by interpolating a higher resolution DEM or by increasing the number of triangles (by recursive subdivision, see Fournier, 1982) before sampling colours from the image. In either case, an image of higher resolution than the DEM should not be sampled due to the possibility of 'aliasing' (e.g. intermittent sampling of a bright feature such as a road). Instead the image resolution should be reduced by area-averaging to around the DEM resolution. Figure 2.3 is an example of a static perspective view of a SPOT panchromatic image shown alongside a black-and-white version of a false-colour composite image derived from LANDSAT-TM.

To resolve hidden surfaces, a number of approaches have been tried. In general, three dimensional surfaces composed of, for example, intersecting opaque and translucent facets require a much more complex rendering schema than that of a DEM (see, Sutherland, 1974). After all, a DEM already contains sorted position information, and for our application, surfaces are known to be opaque and non-intersecting.

Resolving hidden surfaces with depth buffers.

Depth buffering is a simple method for rendering objects, particularly with surfaces composed of planar polygons and triangles into an image. For each image pixel we store the colour, and the range to the object visible in the pixel. Initially the image is blank and the depth of all pixels is set to the largest representable value. Rendering primitives are then scan converted to pixels, with the pixels only rendered on the framebuffer if the range ('z-depth') is less than the previous value.

The renderer for our triangle-based system uses the depth-buffer provided on a special-purpose graphics accelerator board attached to one of our Sun workstations. This device includes hardware to perform 3D transformations using homogeneous co-ordinates (Newman, 1983). However, this device is limited in the accuracy to which it can store range (16 bits), so objects at similar depths projected to the same image pixels may not be rendered in the correct order. Because homogeneous co-ordinates are used, the depth value stored in the buffer is actually a nonlinear increasing function of the true depth, with the effect of increasing depth resolution in the foreground, but lowering resolution in the background. In video animated sequences, this can cause distant objects to appear to 'crawl'.

To solve the problem of aliasing artefacts, the A-buffer (standing for anti-aliased, area averaged, accumulation buffer, see Carpenter, 1984) has been developed. This is a general hidden surface technique that can resolve visibility among a completely arbitary collection of opaque, transparent and intersecting objects, to a resolution many times greater than the traditional z-buffer.

Rendering operations using this system are identical to that of the Z-buffer: modelling primitives are diced to flat polygons which are then thrown at the A-buffer in an arbitary order. Every polygon is resolved after first being clipped to pixel extent. The visibility resolver, using a box filter, maintains a list in z of sub-pixel sizes. When all objects are resolved, we do a weighted summation of any sub-pixel values.

The A-buffer is useful in many areas, such as correctly resolving very small polygons. Its generality may be more than is needed for simple surfaces such as terrain, but for complex self-enclosed 3D objects, it would appear to be ideal.

Ray-tracing.

A numerical modelling scheme has been developed in which physical processes can be simulated. These involve the production of an image on a detector from the scattering of light from a terrain. At the present time, no attempts have been made to include atmospheric effects.

The model was constructed using a simulation technique known as Monte Carlo sampling (see Kajiya, 1986). This allows a finite number of rays to represent a huge number of actual photon (wave) interaction events with matter by ensuring that optimum sampling is done of the critical parameters (e.g. ray direction, time of emission, wavelength and polarisation).

The critical path for each ray starts at the detector where rays are fired through the principal point into the environment. This is done because we know that rays emitted from the detector will contribute to the image whereas most of the rays emitted from the light source will not. After intersection with the terrain, the rays are absorbed and then re-emitted until they either hit another surface or a source of illumination such as the sky or the Sun.

Different surface reflectance models can be represented using the Bi-directional Reflectance Distribution Function (see Nicodemus, 1977) which is a measure of the ratio of exitant radiation to irradiance for all solid angles of incidence and exitance and for all wavelengths.

Monte Carlo sampling uses probability functions to bias the sampling to those parameters which most contribute to the measurement. For example, to reduce sampling errors at the detector, a Poisson distribution is used to jitter randomly the direction of the emitted rays within the sample area (Mitchell, 1987). An analagous technique is used to reduce the sampling of rays with wavelengths where the detector has a poor response and to bias the re-emission of rays from a surface into the direction of the principal plane of illumination.

An example of the application of this technique is shown in figure 2.4 which demonstrates the potential of this technique to produce realistic visualisations of terrain including the effects of sky radiance.

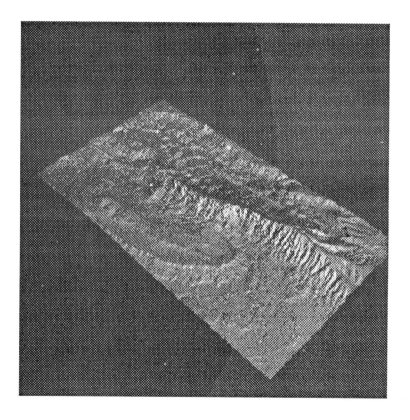

Figure 2.4. Ray traced image of Montagne Sainte Victoire at 8am on 12th May

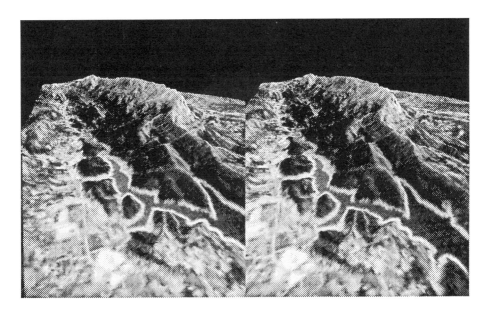

Figure 2.5. Stereo pair (SPOT overlay)

2.5. Display.

There are a number of animation techniques which can be used to display rendered perspective views of terrain.

These display techniques include :
1. Single camera viewpoint;
2. Stereo camera viewpoint;
3. Multiple camera viewpoint (e.g. ahead, from above, at the side);
4. Fixed focus of interest for camera(s);
5. Movable focus of interest for camera(s);
6. Interactive specification of camera(s) flightpath;
7. Pre-defined specification of camera(s) flightpath (e.g. view from a pre-determined flightpath)

In addition to describing the camera viewpoints (i.e. absolute orientation parameters) we can also specify the velocity and acceleration of the camera and the duration of each rendered sequence. From this, the number of frames required for rendering any one piece of animation can be calculated and the animation can be left as a batch task. Finally a description of our equipment for the production of video animations will be made.

Multiple viewpoints.

Single views of landscape can easily be generated using any of the rendering techniques described in the last section. However, their use is limited as the investigation of any single area requires repetitive video animations to be made. A more efficient scheme is to render a number of views concurrently on different workstations (or with transputers, see Muller, 1988a) using the same DEM database, where each collection of views at a single instant of time either shows a stereo view of each scene or multiple views from a number of, possibly, orthogonal viewpoints. This also offers the possibility of focusing interest on particular areas of the landscape by fixing one viewpoint whilst moving the others and varying the focal length of our virtual cameras. These multiple viewpoints can easily be displayed using the NeWS/X11 (Gosling, 1988) window management system.

Stereo views (see example in figure 2.5) have been developed for the analysis and editing of blunders in automatically generated DEMs. They can currently either be viewed using a mirror stereoscope (which limits it to a single observer) or using a variety of anaglyptic (Red/Green) techniques. Future planned developments include 3-D digitising pens or gloves for editing and/or camera viewpoint manipulation and Liquid Crystal Shutter screens placed in front of CRT monitors or video projectors (see, for example, Fergason, 1983) which will permit 24-bit colour images to be displayed stereoscopically.

Animation.

Techniques have been developed to enable the easy interactive specification of the imaginary flight path of our virtual camera(s) around the landscape. Since it is not possible yet to visualise our landscape in real time, we preview the 'flight' with a wireframe model of the terrain, and use this specification to drive the renderers.

Figure 2.6 shows a typical screen with this flight specification software. Notice in this figure that there are a number of distinct functional units including:

1. (at top left) a control panel where camera parameters can be specified and changed;
2. (at bottom left) the flight-path can be drawn superimposed on an intensity-range image of the DEM;
3. (at top right) the vertical profile corresponding to the planimetric path defined in the bottom left can be specified and changed;
4. (at bottom right) a wireframe without hidden line removal which can be previewed at video rates.

An optical mouse attached to the workstation allows us to draw a planimetric flight path over a view of the DEM (or satellite image or both). To

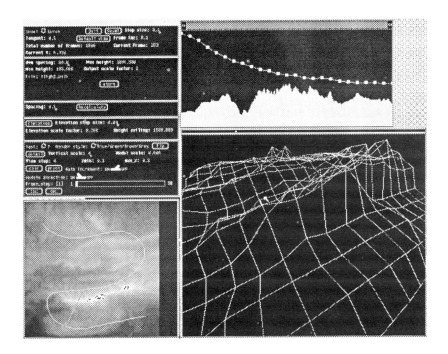

Figure 2.6. Flight path specification

control viewing orientation, two sub-windows within the control panel allow us interactively to specify camera direction. The top window contains a small slider which represents the position of the currently displayed frame in the whole sequence. This can be modified by the left button. The right button contains a menu for the options of locking the displayed frame to the spline control points. Also in this menu is the option to (re-)play the sequence. Figure 2.7 shows an illustration of several views generated from this interactive specification.

In addition to interactive specification, views can be generated for a pre-specified flightpath. This can be used to re-visit a site after modifications have been made to a stereo matching algorithm or rendering technique. In the future this will be linked into our digital mapping system (LITES2 from Laser-Scan Laboratories) to enable "flights" to be made from ground-level along a road and use digitised aerial or ground-based photographs to render the views.

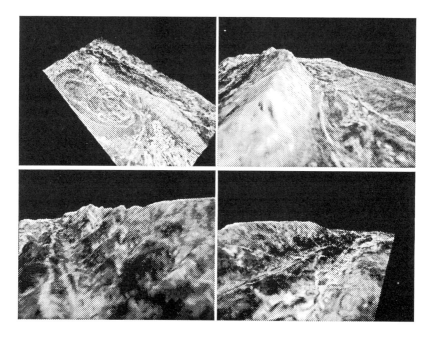

Figure 2.7. Views from various points along flightpath

Hardware.

Figure 2.8 shows a schematic diagram of the hardware being used to produce video animations. The primary means of data transfer is digital and uses our co-axial Ethernet cabling. Data can be transferred from our rendering engine (the GP accelerator) on our Sun-3/160 to the 24-bit Primagraphics CCIR/RS-170 framestore on our Sun-3/260. Data can also be created on the VaxstationII/GPX using the LITES2 or MATRIX packages and transferred using software protocol conversions to our Sun-3/260. Finally data from the two digitising CCD cameras in our Kern DSR-11 can be transferred from the VMS/GPX to our Unix/Sun via the same route.

The Primagraphics hardware contains a device driver which enables the Sun window management system to be used and enables any Sun user on the network to generate and display frames. Finally, the EOS animation controller allows single-frames to be written from the Primagraphics framstore to standard half-inch U-matic videotape from which VHS dubbings can be made (CCIR/PAL only).

Timings for renderings of CCIR $768 \times 575 \times 8$-bit frames are shown in table 2.1. The transfer from one workstation to another is a negligible

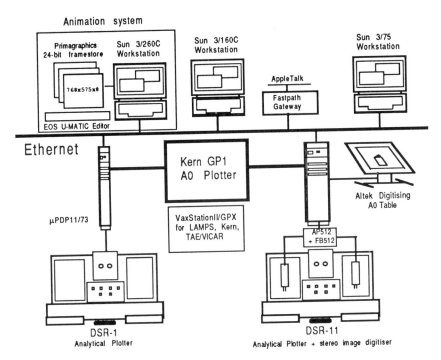

Figure 2.8. Visualisation hardware and data sources

Table 2.1 Rendering Times

Equipment	Shading	No. triangles	Speed
Sun 3/160	Lambertian shading	191730	320 triangles/sec
3/160 + GP1	SPOT image overlay	426560	4080 triangles/sec
Sun 3/260	Ray-traced (32 rays/pixel)	191730	1.5 pixels/sec

fraction of the total time.

2.6. Discussion and conclusions.

Video animated sequences of topographic data open up new possibilities for communicating geographical spatial information which is currently encrypted in paper maps or digital databases. It frees the viewer from two-dimensional degradations of our three-dimensional world and can provide significant aids to specialist photo-interpreters who may be interested in only small areas of a satellite image or only one aspect of the information contained within a satellite image.

Future work will include the incorporation of larger databases (we have already produced a number of animations of the whole planet using a 5' DEM database from NOAA) and complete stereo-matched SPOT scenes using scanline techniques based on Robertson, 1987. Experiments will shortly begin on using a new generation of graphics accelerator (see England, 1988) to speed up rendering times and extensions to arrays of transputer elements (see Muller, 1988a) are currently being designed. It is our firm belief that within a few years commercial systems will be available for doing animations at the pace of the analyst and this will become a standard technique for the analysis of satellite data and the output of image understanding systems.

Acknowledgements.

Small sections of the satellite image renderer were written under an SERC grant concerned with the development of "real-time 2.5D Vision Systems" (see Muller, 1988c). The ray-tracer was developed under a research contract with BP Petroleum Development Limited. We would like to thank Lyn Quam (SRI), Jeffrey Hall (JPL) and Alex Pentland (MIT) for useful discussions and M. P Foin, IGN for providing the SPOT data, underflight photographs and control points through the aegis of the SPOT-PEPS campaign.

References.

Bhanu B. and Ho K., CAD-based 3D object representation for robot vision. *IEEE Computer*, 22–36, August.

Blinn J. F., 1977, Models of light reflection for computer syntheiszed pictures. *Computer Graphics (Proc. SIGGRAPH 77)*, 11(2), 192–198, 1977.

Brew A. N. and Neyland H. M., 1980, Aerial photography. In *Manual of Photogrammetry*, pages 280–283, ASPRS, 1980.

Carpenter L., 1984, The A-buffer, an antialised hidden surface method. *Computer Graphics (Proc.SIGGRAPH '84)*, 18(3).

Cook R. L. and Torrance K. E., 1981, A reflectance model for computer graphics. *ACM Trans. on Graphics*, 1(1), 7–24, 1981.

Day T. and Muller J-P., 1988, Quality assessment of digital elevation models produced by automatic stereo matchers from SPOT image pairs. *IAPRS*, 27–

A3.

Dowman I. J. and Muller J-P., 1986: Real-time photogrammetric input versus digitised maps: accuracy, timeliness and cost. *Proc. Auto-Carto London*, 1, 583–543.

England N., 1988: Application acceleration: development of the TACC-1. *Sun-Technology*, 34–41, Winter.

Fergason J. L., 1983, *Liquid Crystal Display with Improved Angle of View and Response Times*. Technical Report 4,385,806, US Patent, May.

Fournier A., Fussell D. and Carpenter L., 1982, Computer rendering of stochastic models. *Comm. ACM*, 25(6):371–384.

Frenkel K. A., 1988, The art and science of visualising data. *Comm. of the ACM*, 31(2):111–121.

Garvey T. D., 1987, Evidential reasoning for geographic evaluation for helicopter route planning. *IEEE Trans. on Geoscience and Remote Sensing*, GE-25(3): 294–304.

Gelberg L. M. and Stephenson T. P., 1987, Supercomputing and graphics in the earth and planetary sciences. *IEEE CG & A*, 26–33, July.

Gosling J. Hoeber T. and Rosenthal D., 1988, Programming with news. *SunTechnology*, Winter, 54–59.

Gourard H., 1971, Continuous shading of curved surfaces. *IEEE Trans. Computers*, C-20(6), 623–629, 1971.

Kajiya J., 1986, The rendering equation. *ACM SIGGRAPH86, vol.20(4)*, 6.

Kolbusz J., 1988, *The Visual Representation of 3D Scientific Data*. Bsc Project Report, Dept. of Computer Science, University College London, March.

Magnenat-Thalmann N. and Thalmann D., 1987, An indexed bibliography on image synthesis. *IEEE CG & A*, 27–38, August.

McCormick B. H. DeFanti T. A. and Brown M. D., 1987, Visualisation in scientific computing. *Computer Graphics*, 21(6):6–26.

McCullagh M. J. and Ross C. G., 1980, Delauney triangulation of a random data set for isarithmic mapping. *Cartographic Journal*, 17(2):93–99.

Mitchell D., 1987, Generating antialiased images at low sampling densities. *Computer Graphics (Proc. SIGGRAPH87)*, 21(4).

Muller J-P., 1988a: Key issues in image understanding in remote sensing. *Phil. Trans. R. Soc. Lond. A*, 324:381–395.

Muller J-P., 1988b, Image understanding system components in remote sensing. In *Proc. ICPR, Beijing, China*, October.

Muller J-P. and Anthony A., 1987, Synergistic ranging systems for remote inspection of industrial objects. In *Proceedings of "2nd Industrial and Engineering Survey Conference"*, London.

Muller J-P. and Saksono T., 1988, Fractal properties of terrain. *IAPRS*, 27-A3.

Muller J-P. Collins K. A. Otto G. P. and Roberts J. B. G., 1988, Stereo matching using transputer arrays. In *Proceedings of the XVIth International Congress of ISPRS, Kyoto, Japan, IAPRS 27-A3*.

Newman W. M. and Sproull R. F., 1973, *Principles of Interactive Computer Graphics*. McGraw-Hill, New York.

Nicodemus *et al.*, 1977, *Geometrical considerations & nomenclature for reflectance*, NBS Monograph 160, US. Dept. of Commerce.

O'Neill M. A. and Dowman I. J., 1988, The generation of epipolar synthetic stereo mates for SPOT images using a DEM. In *Proceedings of the XVIth International Congress of ISPRS, Kyoto, Japan, IAPRS 27-A2*.

Phong B-T., 1975, Illumination for computer-generated pictures. *Comm. of the ACM*, 18(6), 311–317.

Quam L., 1987, private communication.

Robertson P. K., 1987, Fast perspective views of images using one-dimensional operations. *IEEE CG & A*, 47–56, February 1987.

Rogers D. F., *Procedural Elements for Computer Graphics*. McGraw-Hill, New York, 1985.

Schachter B., 1983, editor, *Computer Image Generation*. John Wiley & Sons, New York.

Sutherland I. E. Sproull R. F. and Schumacker R. A., 1974, A characterisation of ten hidden surface algorithms. *ACM Computing Surveys*, 6(1), 1–55.

Winkler *et al.*, 1987, A numerical laboratory. *Physics Today*, 28–37, October.

Wolff R., 1987, Panel on "the physical simulation and visual representation of natural phenomena". *Computer Graphics (SIGGRAPH 87)*, 21(4), 337–338.

Woodham R. J., 1985, A computational vision approach to remote sensing. In *Proc. IEEE Computer Vision & Pattern Recognition Conf.*, pages 2–11.

Zyda *et al.*, 1988, Flight simulators for under $100,000. *IEEE CG & A*, 19–27, January.

3
The Transportable Applications Environment —an interactive design-to-production development system

Dorothy C. Perkins, David R. Howell and Martha R. Szczur
Data Systems Technology Division, Code 520
NASA Goddard Space Flight Center
Greenbelt, Maryland 20771
ARPANET: Howell%DSTL86.SPAN@DFTNIC.GSFC.NASA.GOV

3.1. Overview.

The Transportable Applications Environment (TAE) is an executive program that binds a system of application programs into a single, easily operated whole, and supports user operation of programs through a consistent, friendly and flexible interactive user interface. TAE was developed to better serve the needs of the end user, the application programmer, and the system designer. This chapter describes the philosophy and architecture of TAE.

Many computer solutions to analysis or information management problems use multiple programs to perform their tasks. For some situations, the programs are numerous, complex, and interrelated. Thus, both the system developer and the end user can benefit from help in organising and managing the various operations. As an executive, TAE simplifies the job of a system developer by providing a stable framework on which the system can be built, thereby decreasing the time between conceptualization and first implementation. Moreover, TAE integrates the activities of the system and cooperates with the host operating system to perform functions such as task scheduling and I/O.

TAE also supplies the interface between the aggregate of application programs in a particular computer system and the user. The original TAE human-computer interface, designed for a standard ASCII terminal, supported command and menu interfaces, information displays, parameter prompting, error reporting and online help. Recent extensions to TAE include support for modern graphic workstations with a window-based, modeless user interface.

History of TAE.

A short history of the system and of the policy decisions made during development gives an idea of how the TAE designers addressed the needs of system developers and users.

The National Aeronautics and Space Administration (NASA) funds scientific research in a number of fields. Most of these activities rely heavily on modern data processing techniques. In fact, most scientists and engineers—in fields such as meteorology, aerodynamics, climatology, oceanography, astronomy and astrophysics—require sophisticated data analysis packages. In 1975, NASA's Goddard Space Flight Center built an interactive analysis system for research scientists in meteorology (Bracken et al., 1977). This was followed in 1977 by a second system for studying atmospheres (Dalton et al., 1981), and in 1978 by yet a third for the interactive display and correlation of United States census data (Dalton et al., 1979).

Planning began in 1979–1980 on four other systems supporting meteorology, oceanography, earth resources studies, and data base management. Although the planned systems had unique application software, their system requirements overlapped in many features and functions. It became obvious that by separating out the applications, there remained a common core of structures, system service routines and user dialogue that could be designed to serve all the application software. In addition, limited manpower, coupled with the enormous resources needed to develop and to maintain all these systems simultaneously, led to the decision to design a common application executive to serve the new systems.

From the beginning, the designers' major concern was for the user community. The reusable executive made sense because, despite some diversity, the potential user communities all made essentially the same generic demands on their supporting computer systems. These common user expectations determined the initial requirements for the executive:

1. a consistent, controlled, interactive, and easily learned interface;
2. easy incorporation of interactive image processing and graphics;
3. batch processing; and
4. integration of potentially large collections of application programs.

Experience with prior systems and awareness of the rapid advances in technology helped the designers to further refine these requirements.

• The executive had to be usable by novice and casual users, yet give flexibility and freedom to expert users.

• Users should be able to locate programs easily and access online information explaining substance and operation of the system.

• The executive must shield the end user from the host operating system.

• The executive must make the system easy to reconfigure: new programs should be added with ease.

• The executive must be portable to other machines.

- The executive must supply common services needed by application programs.
- The system must be independent of discipline or data.

At TAE's inception, all of these attributes had not been incorporated into any existing system. Even though the industry recognized the efficacy of general-purpose, versatile, and reusable software, by 1979 few application executives were operational. The increasing interest in application executives over the past few years can be explained by the rapidly escalating cost associated with software development.

In TAE's conceptual phase (1979–1980), three research areas exerted a strong influence on the designers' thinking. The virtual operating system designed at the Lawrence Berkeley Laboratory (Hall *et al.*, 1980) gave insight into the problem of imposing uniformity on diverse computers so that programs are portable. The work then being done in human-computer interfaces provided information on questions of system versatility and user satisfaction (Shneiderman, 1980; Moran, 1981; Demers, 1981). The third influence came from research on the kinds of design tradeoffs necessary to foster a suitable environment for an extensible, portable interactive system (Ling, 1980).

TAE has grown from a limited prototype (first made available in August, 1981) to a widely used, mature system. It has undergone an evolutionary growth, shaped by the experience and critiques of its users. At every stage in development, user comments were incorporated into revisions. In addition, each phase of development and enhancement has taken into account advances in virtual operating systems, human factors research, command language design, standardisation efforts, and software portability. TAE is independent of discipline and data type, and is being used

- for scientific analysis;
- for image processing;
- as a front end for large or experimental hardware;
- in the university classroom environment;
- for rapid system prototyping;
- in software engineering laboratories; and
- for business and industrial applications.

3.2. TAE features.

TAE is now in its second major development phase. The first phase, called "TAE Classic," was built for use with interactive alphanumeric terminals. It is operational and is widely used. The second phase, "TAE Plus," is an enhancement to TAE Classic that allows modeless interaction in a window-based, graphical environment. At the time of this publication, it was installed in prototype form at several test sites. It retains all the functionality of, and is upwardly compatible with, TAE Classic.

Central to the TAE concept are two basic elements: procs and parameters. A proc is some function such as data analysis, data display, or housekeeping, which a user wants to execute. Internally, a proc may be either a process (an executable program) or a command procedure (a predefined sequence of TAE commands including the execution of procs). A proc is made known to TAE by the existence of an editable text file called a proc definition file (PDF), which internally identifies itself as a process or procedure. While TAE understands the distinction between proc types, to the end user they are treated identically.

Each proc has associated parameters defined within the proc's PDF. By setting or selecting parameter values, the user adapts the action of a proc to immediate needs. For example, the data file on which a proc is to work may be an input parameter. Parameters may have default values, multiple values and restricted values, (including enumerated lists), that will be used by TAE to display a value list for user selection or to validate user input. Parameters may be required or optional. Parameter sets may be saved for repeated use.

Sets of related procs are stored in "libraries". A single system can have many libraries of procs, including user-defined libraries. Users can employ all libraries or a selected subset. To locate a specific proc, TAE searches the libraries in order from user libraries through application-specific libraries to system-wide libraries.

TAE consists of six components, as follows:

1. The TAE Classic "TAE Monitor" (TM) handles all user-computer communications in TAE Classic. It supports two modes, "menu" and "command," collects and packages user-selected parameter values for use by procs, and manages the execution of application programs. It contains a command interpreter for TAE command procedures. TAE Classic user interactions are supported on standard ASCII terminals and on window-based, mouse and keyboard driven graphic terminals.

2. The TAE Plus graphical interaction capabilities, which allow the definition and execution of "modeless" user interactions on high resolution graphic workstations.

3. The TAE subroutine library, which provides several packages of commonly needed services for application programs.

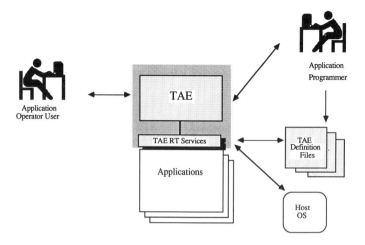

STRUCTURE IN TAE - BASED SYSTEM

Figure 3.1. Structure in TAE-based System

4. The TAE Plus "WorkBench," which allows application interface developers to design display screens and make them known to TAE for subsequent execution.

5. A TAE subsystem for management and control of raster imaging devices.

6. A TAE subsystem for interfacing among TAE systems across a communications network.

Each will be described in turn.

TAE Classic User Communication through TM.

The user "sees" TAE Classic as an interface having two modes: menu and command. Novice or casual users make their way through the system of menus, augmented by extensive online help. Experienced users have a powerful language for commanding the system and controlling the environment. Both menu and command users have online explanation of the system. In either mode, the user's ultimate purpose is the same: to execute the application functions of the system. Figure 3.1 depicts the user–system communication through the TAE monitor.

TAE Classic Menu Mode.

The menu user sees a controlled, ordered interface in which he or she chooses a path through the system from a series of formatted option lists. The choices which can be made from any one menu are restricted, but menus are typically designed to group related functions, in anticipation of the likely choices a user will make. A TAE system arranges menus in a tree; the leaves of the tree are procs. Any one menu or proc may appear in the menu tree as often as desired. Menus are defined by disk-resident text files.

In menu mode, TAE communicates with the user by means of four types of displays: (1) menu, (2) tutor, (3) help, and (4) message. All four show two different kinds of material:

- Content: the instructions, information, or data the user needs at that specific time.
- Prompt-line Options: a set of choices allowing the user to interact with the display.

Menu.

A TAE menu is normally a CRT display containing entries and menu prompt-line options. Figure 3.2 shows a simple TAE menu for an ASCII terminal. Figure 3.3 shows the same menu for a graphic terminal.

Tutor.

Once a user reaches a proc, he or she is "tutored" to enter the desired parameter values. In a tutor session, the user sees a screen containing a formatted list of parameter names and either their previously established values or their defaulted values. With a tutor display, users may set new values or edit old values. Users may also save sets of parameter values for future restoration. Figures 3.4 and 3.5 show a typical tutor screen for ASCII and graphic terminals.

The name "tutor" implies part of the function of this mode—to teach a user the command language of TAE. The tutor display assists a user in understanding and entering parameter values. "No Screen," a variation on tutoring which does not require a full screen and minimises the amount of information written to the terminal, is also available.

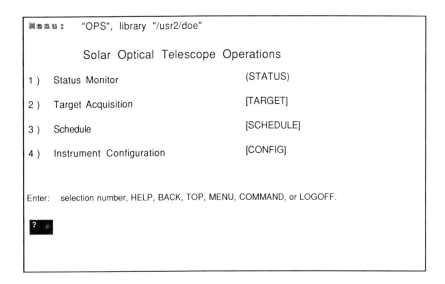

Figure 3.2. A TAE Menu for an alphanumeric terminal

Help.

To be an effective interface, an executive must be self-contained. That is, it should be able to inform the user of its resources and give instructions for their use. In order to be as self-teaching as possible, TAE provides the user with help on five categories:
- a proc, TAE command, or topic chosen by system implementors; this help covers system functionality;
- a menu;
- the operation of a mode (menu, command, tutor);
- the parameters of a proc; and
- a TAE message.

A typical TAE help display may have several screens, each of which contains context-specific explanatory information and standard help options.

TAE is designed so that the help text may be timely and useful. In addition, the same help text can be accessed in several different ways. Such redundancy is a planned feature of TAE; it ensures that the user can reach required information from several directions, using what knowledge of the system he or she has available at the time.

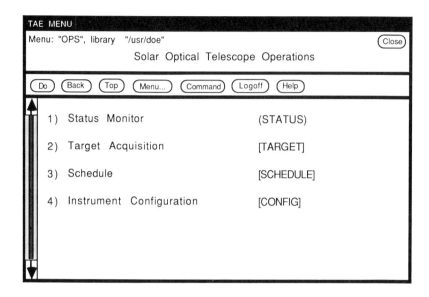

Figure 3.3. A TAE Menu for a graphic terminal

Messages.

A message display is used to report an error or a piece of information to a user. It consists of a single comment line that appears on the current display screen. The user may then choose from the options appearing on the prompt line of that particular screen. A typical message has the form:

[WHOSE-WHAT] description

where WHOSE identifies the source of the message and WHAT is a key. Any message may be supplemented by additional help, which a user accesses by typing "?" when the message occurs.

TAE Classic Command Mode.

Users who employ the TAE Command Language (TCL) can freely direct system activities. Unlike menu users, who have a limited set of possible actions and very little to remember at any one time, TCL users are bound only by the breadth of the system and their ability to recall how to use the various resources.

Like menu users, command users are primarily interested in executing the application and utility procs of a system. In a simple command

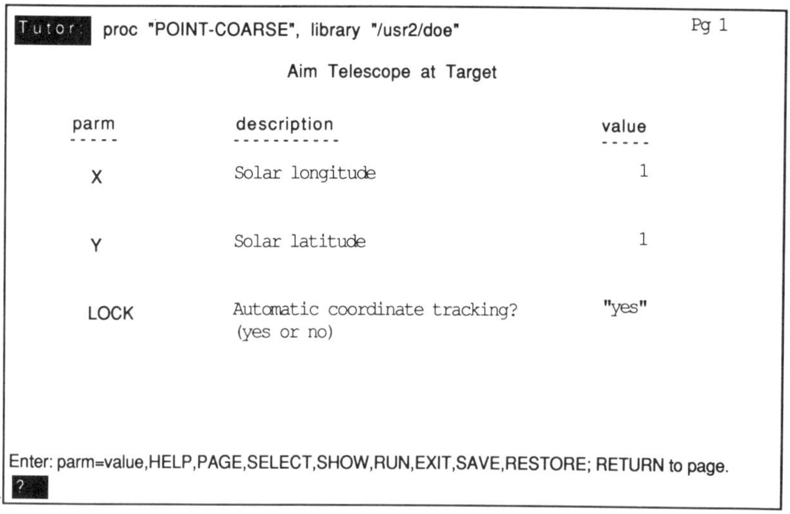

TAE—an interactive design-to-production development system

```
Tutor:  proc "POINT-COARSE", library "/usr2/doe"          Pg 1
                     Aim Telescope at Target

        parm         description                    value
        ----         -----------                    -----
         X           Solar longitude                  1

         Y           Solar latitude                   1

        LOCK         Automatic coordinate tracking?  "yes"
                     (yes or no)


Enter: parm=value,HELP,PAGE,SELECT,SHOW,RUN,EXIT,SAVE,RESTORE; RETURN to page.
 ?
```

Figure 3.4. A Tutor Screen for an alphanumeric terminal

```
TAE TUTOR
TUTOR: proc "POINT-COARSE", library "/usr/doe"              (Close)
                     Aim Telescope at Target

(Run) (Show...) (Initial) (Save...) (Restore...) (Structure) (Exit) (Help)

         parm          description                      value

          X            Solar longitude                    1

          Y            Solar latitude                     1

         LOCK          Automatic coordinate tracking?   "yes"
                       (yes or no)
```

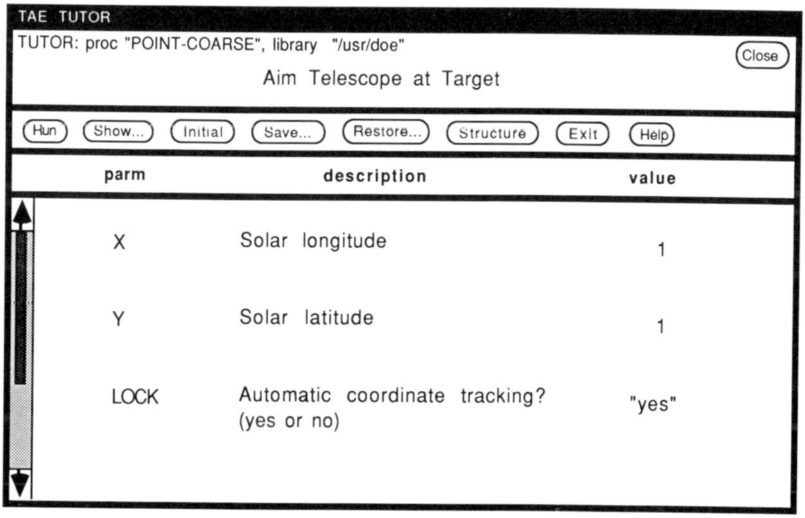

Figure 3.5. A Tutor Screen for an alphanumeric terminal

statement, the user enters the name of a proc and parameter values, either positionally or by keyword. Figure 3.6 gives a sample sequence of commands.

```
TAE>  iatstat
TAE>  alloc chinook
TAE>  totv in=hall1 out=allen1 type=bw
TAE>  totv in=hall2 ou=allen2 type=bw
TAE>  totv hall3 allen3 bw
TAE>   defcmd lp "loop (allen1,allen2,allen3)"
TAE>  lp
TAE>  stoploop
TAE>  view allen1
TAE>  vsprint |runtype = batch| allen1
TAE>  tu overlay
```

Figure 3.6. Sample TAE Command Sequence

Command users also have access to help and messages. In addition, they may tutor on a proc or switch to menu mode at any time.

Users may define their own commands by assigning an alias to a command string. Whenever that alias is used as a command, TAE substitutes the equivalenced string in place of the alias before processing the command.

Users may also set up standard sequences of commands by creating command procedures—collections of TAE commands, executed as a single, named function. For this purpose, the language contains control directives (e.g., LOOP, IF-THEN-ELSE) typically found in a procedural language. Other features such as global and local variables, substitution, assignments, input and output parameters and expression evaluation support programming through procedures. All TAE PDFs are written in the TAE command language.

A simple in-line command editor allows users to recall and edit or resubmit previous commands.

TAE Plus User Interface.

TAE Plus, developed to open TAE to use on modern workstations, enables graphics-oriented applications to easily utilize workstation features such as multiple windows, graphical icons and objects, colour and various input devices. It supports the simultaneous execution of application procs in multiple windows and the use of multiple windows for a single proc. Unlike TAE Classic, the design of the screen content is not controlled by TAE, although TAE provides default layouts for both windows and screens. Figure 3.7 depicts some of the kinds of interactions supported in TAE Plus. The values of the objects are taken from the parameter validation lists stored in TAE PDFs (proc definition files). Figure 3.8 depicts an example screen developed in TAE Plus. The TAE Plus user interface utilizes features of TAE Classic such as procedure interpretation, help display and parameter packaging.

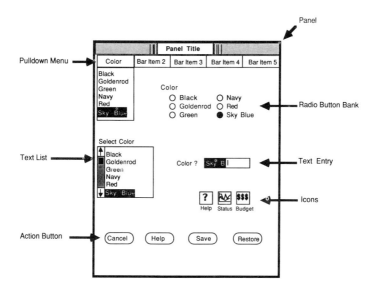

Figure 3.7. TAE Plus Interaction Objects

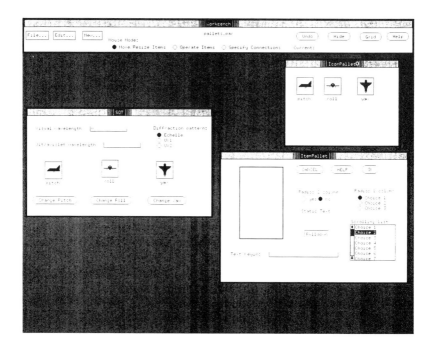

Figure 3.8. Example TAE Plus Screen

TAE Subroutine Library.

The TAE Subroutine library provides commonly needed services to application programs. The routines also isolate application code from host operating system services. Five packages of subroutines are supported.

1. Parameter Processing accepts user-selected parameter values, validates them, and passes them to the associated proc.

2. Window Programming Tools control the TAE Plus user interface. These subroutines define, display, receive information from and delete TAE Plus windows, and TAE Plus interactive display items, called interaction objects.

3. Image I/O handles band sequential or band interleaved images and allows user-defined labels to be inserted at the front of an image file. This package is optimised for I/O efficiency.

4. Terminal I/O gives an application simple alphanumeric I/O with carriage control, cursor control and screen clear features.

5. Message Display gives programs access to TAE message handling, thereby allowing consistency across all messages, whether generated by

TAE or by applications.

TAE Plus WorkBench.

The TAE Plus WorkBench is a development tool that allows the designer of an application's user interface to interactively define, position and set the display attributes of panels and interactive objects. These displays are saved by the WorkBench, and can then be accessed directly by application programs using window programming tool subroutines. Figure 3.9 depicts the structure of TAE Plus with the WorkBench.

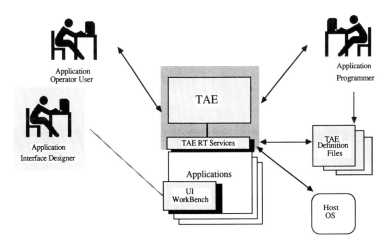

Figure 3.9. TAE Structure with WorkBench

Display Management Subsystem.

The Display Management Subsystem (DMS) provides standard, device-independent access to raster image display devices as well as user services to support this access. This subsystem centers on two concepts, device categories and "images."

Device categories identify abstract imaging devices. Specific hardware is identified not as a particular model or type but by capacities and capabilities. For example, one device might be defined as having three refresh memories, three lookup tables, one cursor, one monitor and one overlay plane. This device, used to capacity by an application program, could display a full colour image with overlaid graphics, and would allow selection of points using the cursor. Programs work with the abstraction of a device, and are linked to specific hardware only at run time. Thus, on one computer system an application program can be set up to work with more than one brand or type of display device.

The "image" concept captures the view of data which a user has on the raster device and permits the user to give that view a name. Any image which a user is viewing dynamically is composed of various hardware features configured in a certain pattern by the user. For example, a full colour image requires three refresh memories, each passed through its own lookup table, and all three routed to a colour monitor. A multiple-memory device could have several such images stored on it. A user may name each such entity, and recall the entire configuration by that name at a later time. This allows the user to focus on the data and not the display device.

DMS has a modular design which supports portability among devices and facilitates both hardware and software expansion. As Figure 3.10 illustrates, users invoke their specific display applications—TAE procs—from a workstation terminal. The application programs use a set of generic service routines for all image device access. Two sub-packages, which may be modified without affecting the generic routines, are table access routines and dévice-dependent routines. Only these lowest levels must be modified to accommodate each type of display device on a system. (For more on DMS, see Perkins *et al.*, 1984).

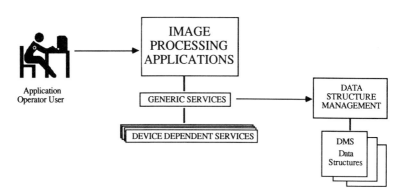

Figure 3.10. Display Management Subsystem Components

Remote Communications Job Manager.

The Remote Communications Job Manager (RCJM) provides TAE users with a uniform interface to a network. RCJM allows a user on a local machine to initiate and control a TAE proc on a remote machine. A user may tutor on a remote proc, access help on a remote proc, invoke a remote proc and copy text or saved TAE parameter files between the user's machine and a remote node.

RCJM provides the end-user interface layer of a distributed system and calls on a network support package, either DECNet or TCP/IP, to complete the network communication.

3.3. Building a System with TAE.

TAE is not a complete system in itself. Rather, it provides a core of services needed in any interactive system, and can be used as the framework upon which customized applications can be installed and managed. The major TAE components—user interface, command language, proc activation, and service subroutines and packages—have been described earlier. This section presents some ways in which a system builder can construct a TAE-based system to suit a specific situation.

Structuring A User's Environment.

How a user's environment is defined and controlled is critical to a system's success. For the most part, a user views an integrated system as a tool and wants to accomplish a task with as little intrusion as possible. Design choices can determine the appearance of the system, ease of use, and the smoothness of flow from one function to the next. Three aspects of TAE—global variables, system logon and logoff, and the formatting and sequencing of text frames—help the system builder to shield the user from complicated interfaces, as well as aiding in adapting an interface for special needs.

Global Variables.

Globals are session-wide variables used to pass information between procs and to supply the TAE Monitor with dynamic control information. In a specific system, certain parameters or conditions—such as file names, area definitions, or an index into a database—may appear repeatedly. To avoid requiring the user to re-enter the information, the parameter may be set up as a global variable.

Logon/Logoff.

At the beginning of a session, TAE executes a system-wide logon procedure which can perform a variety of functions, such as setting up standard application libraries and global variables, or restoring variables saved in a previous session. An installation "bulletin board" may also be displayed at logon. After the session is initialised, a user logon proc is run to customise the session for an individual user. For example, special application libraries could be added to the standard set.

At the end of a user's session, system and user logoff procs run. These are used primarily for cleanup operations and to save the user's environment.

System Appearance.

The physical appearance of TAE helps the user learn the system quickly. Such attributes as standardised display formats, rigorous categorisation, sequential task subdivisions, and functional redundancy create a consistent learning environment. For example, menu interaction gives an introduction to the system and helps the user construct a mental picture of the system's configuration. Furthermore, menu choices may appear in several places, allowing frequently used entries to be displayed where they logically belong. In adapting TAE for a specific site, care must be taken to produce good menu or screen structures. However, once this conceptualization is complete, its implementation is relatively easy.

TAE provides mechanisms for establishing and formatting extensive online help and message information. Content, however, is particular to each system. Experience has shown that, unguided, programmers will typically write help information for applications in programming terms rather than in user terms. Experience has also shown that poor help information can frustrate and discourage a user, even in an otherwise well-designed system.

Some research has been done in the sequencing and layout of text information for users (Marcus, 1982), and in effective presentation of online information (Carlson, 1983). The latter document is distributed with the TAE installation kit. Further useful comments on communicating with a user can be found in Shneiderman (1980).

Host Commands.

TAE does not replace the host computer's operating system; it overlays this interface with an environment configured for a specific analysis system. In this way, TAE insulates the user from the host's command language and error messages. However, both the terminal user and the software developer may want to use certain functions residing in the host. Rather than reinvent these commands, TAE provides for direct execution of host commands.

Host commands can be executed interactively, defined in a TAE alias command or "captured" in a proc to isolate the user from them. Procs containing host commands may be used in the same way as any other proc. Entire menus of host commands may be created with no applications code present. Using host commands, however, involves a trade-off. While the practice gives more power and flexibility to the system builder, procs containing host commands are not portable.

Prototyping.

A particularly valuable attribute of TAE is the ability to prototype a system. An entire system model can be built without writing a single line of application code. For example, TAE Plus panels and interaction objects, and therefore application screens, can be designed and displayed using only the WorkBench and the TAE command language. These, along with help files, global variables, menus and dummy procs can all be created to demonstrate a new system's planned structure, appearance and capabilities. Users can interact with the system easily and quickly, changing or reconfiguring it by editing text files or using the WorkBench. Such pre-release field-testing of application systems promotes experimentation in appearance and operation of the system. This capability allows early user involvement, encourages meaningful communications between users and developers, and increases the likelihood of user satisfaction with the delivered system.

Modularity and Extensibility.

TAE is an open system, meaning that application programs can be added easily to augment the existing complement of functions. The executive incorporates those qualities of modularity and extensibility that make software as general as possible. TAE allows growth, incremental development, and addition of new functions through the easy installation of new procs. Both procedures and processes may be added dynamically at any time by simply creating a PDF file in one of the libraries, including the user's private library. A menu can be updated quickly by editing the appropriate text file. The system does not have to go through a "sysgen" process to make these changes.

Portability.

Portable code makes up approximately 85% of TAE and all system dependencies are isolated from the portable code. The remaining code is host specific. For this small percentage—less than one-sixth of the total—all or part of each routine must be reimplemented on each new host.

Experience with TAE indicates that a typical minicomputer porting effort requires 4 to 5 person-months. TAE does not address portability problems of application programs brought about by incompatibilities in compilers, word length, or address space. However, most applications can be ported between heterogeneous TAE systems if the application code adheres to a standard (such as FORTRAN 77), uses TAE service routines, and avoids operating system interactions. Figure 3.11 shows systems on which TAE was installed at time of this publication. TAE Plus uses the M.I.T. X Window System (Scheifler, 1987)

TAE Computer Installations

	TAE Classic		TAE Plus Prototype
Current Versions	VAX/VMS VAX/UNIX SUN/UNIX AT&T/UNIX ISI/UNIX Alliant/UNIX IBM RT/UNIX Macintosh/MACWorkstation	Gould/UNIX Apollo/UNIX HP9000/UNIX Jupiter/UNIX CDC/UNIX Altos/UNIX	SUN/UNIX Apollo/UNIX MAC II/AUX VAXStation II/ULTRIX
Ports in Progress or Proposed	IBM/PCDOS IBM/VS PRIME/PRIMOS CDC/NOS DG/UNIX Masscomp/UNIX		VAXStation II/VMS IBM RT/UNIX HP/UNIX

Figure 3.11. TAE host systems

Layering.

Because TAE is a general purpose control program, the system must accommodate users as diverse as data processing clerks, scientists and engineers, business executives, application programmers, system operators, and service personnel. The designers have addressed this requisite by allowing for layering (or subsetting) of the system's functional capabilities and user interfaces to serve the needs of a particular user or application category.

For example, functions and interfaces for an end user may be layered in three different ways:

1. Creating unique utilities and application processes.

2. Using the TAE command language to create higher level or special-purpose procs that either do a larger, more complex task or provide a simpler interface that hides the existence of some parameters from a user.

3. Creating new TAE commands. These would probably be functions which must respond very quickly or special operations on particular kinds of data.

Layering can also be done for the application programmer. The TAE Display Management Subsystem is an example of a layered subroutine package that works with TAE and provides a higher-level programmer interface to display devices.

3.4. Using a System Built with TAE.

The previous section discussed methods which system developers have to establish supportive user environments. This section considers how a user explores and learns the TAE system.

Environment Discovery.

Complex systems which require users to move up and down through levels and to make a series of interconnected choices in the problem-solving sequence have inherent difficulties for human operators. Not having a conceptual overview of the system and losing one's place in a session are frequent complaints voiced by users.

In TAE menu mode, the menus themselves, augmented by online help and tutoring, act as a guide through the system. In TAE Plus, graphical interaction options are visible to a user and can be freely explored. In command mode, the following online system information serves as "navigational aids:"

MENUTREE: Generates a graphic display of all or part of the menu tree.

SHOW: Displays the proc libraries in the order of search procs.

PROCS: Lists all procs in any or all libraries by name with a brief description of function.

HELP: Presents information about system functions and parameters.

DISPLAY: Lists the values of any or all global or local variables.

Environment Modification.

Designers of human/computer interfaces pay particular attention to the sharing of control and responsibility between the human and the computer. A new or infrequent user prefers limited responsibility and a rigidly defined agenda for operating the system. The expert, on the other hand, wants more control over the operating environment. More importantly, a system should be capable of being tailored by a user: as a user gains expertise, he or she should also gain more control.

Some ways in which users control their environment include choosing proc libraries, changing global values that alter the environment, and establishing parameter values for procs. Frequently selected parameter sets can be saved, and restored on the command line or in tutor mode when needed.

Session Log.

Some users may choose to keep a record of all or part of a session. A command to TAE will cause a log to be created. Thereafter, a text file of all user transactions will be generated, including lists of all parameters, messages, and proc termination conditions. Various kinds of formatting of the log for user viewing are available. The log may also be edited to create a new proc that can be executed.

Documentation.

A full library of paper documentation exists for TAE. As of this publication, the library included:
- Conceptual Design
- User's Reference Manual
- Application Programmer's Reference Manual
- C Programmer's Reference Manual
- System Manager's Guide (VAX and UNIX)
- Utilities Reference Manual
- Guidelines for Designing Menus and Help Files
- Primer
- Functional Specification
- System Internals
- Display Management Subsystem
    - Functional Specification
    - Application Programmer's Reference Manual
    - System Programmer's Reference Manual
    - Internals Guide
    - Application Functions User's Guide
- Introduction to TAE Plus, Version 3.1

- Window Programming Tool Programmer's Reference Manual
- TAE Plus Style Guide
- Release Notes
    - TAE Version 2.1 UNIX Implementation
    - TAE Version 2.2A, VAX/VMS Implementation
    - TAE Plus Prototype Version 3.1, Volume I and II

The user's, programmer's, system programmer's, and utilities manuals are included as text files on the TAE delivery tape.

TAE Support Office.

The TAE Support Office (TSO) provides basic information and assists users and developers with specific problems. The TSO staff give tutorials on two levels: (1) an overview of the basic capabilities for the terminal user and (2) an orientation for programmers involved in producing software under TAE. TSO also receives problem reports from TAE sites, investigates the situation, and prepares responses. Additionally, the support team distributes documentation, helps to organise the annual TAE User's Conference and issues a TAE newsletter three times a year. TSO can be reached at the following addresses:

Goddard Space Flight Center
TAE Support Office
Code 522
Greenbelt, MD 20771
(301) 286-6034
FTS 888-6034
Telemail from a NASA facility: [TAESO/GSFCMAIL]GSFC
Telemail outside NASA facility: [TAESO/GSFCMAIL]GSFC/USA
SPAN 6162::TAESO
ARPANET: taeso%DSTL86.span@dftnic.gsfc.nasa.gov

TAE DISTRIBUTION.

TAE (VAX/VMS and UNIX 4.2BSD) is distributed through the Computer Software Management and Information Center (COSMIC) at the University of Georgia, Suite 112, Barrow Hall, Athens, Georgia 30602 (telephone number: (404) 542-3265).

3.5. TAE Utilisation.

When TAE was delivered as a prototype in late 1981, it was incorporated into three new systems at Goddard Space Flight Center (LANDSAT Assessment System (LAS), General Meteorology Package (GEMPAK) and Pilot Climate Data System (PCDS)). As these systems became available to other government agencies and universities, the use of TAE for other applications began to increase. Some of the types and examples of applications which use TAE (as of this publication) are presented here.

Scientific Analysis.

The General Meteorology Package (GEMPAK) is a graphics application which provides gridding and analysis of surface and upper air observations. GEMPAK was originally developed for GSFC's staff meteorologists, but has gained popularity among the university meteorological community, where its use is growing rapidly.

The Atmospheric and Oceanographic Information Processing System (AOIPS) provides a meteorology image display and navigation system which supports research on severe storms by GSFC scientists through the extraction and analysis of wind field parameters from satellite and aircraft data. Functions include radar and stereo display, as well as stereo cloud height analysis tools. AOIPS uses the TAE Display Management Subsystem.

The Interactive Research Imaging System (IRIS) is a collection of meteorology applications used to process weather satellite imagery for research programs at Colorado State University.

Image Processing Applications.

The Land Analysis System (LAS) is an enhanced version of the Landsat Assessment System. It provides general image analysis and Thematic Mapper data processing capabilities. As public domain software, the LAS has been distributed to several government agencies and universities. LAS uses the Display Management Subsystem.

The Multimission Image Processing Laboratory (MIPL) is a general image analysis system which provides planetary image processing. This system was developed by the Jet Propulsion Laboratory (JPL) and is used at numerous universities and research laboratories which have a cooperative program with JPL.

Data Base Management.

The Climate Data System, part of the NASA Space Science Data Center, is a GSFC pilot system for managing scientific satellite data and limited surface data. The Climate Data System allows users to identify, access, manipulate and display a number of weather, climate, atmospheres and oceans data sets. The Climate Data System uses TAE for its user interface and as a front end to a commercial data base management system and a commercial graphics package. The Planetary Data System (PDS), located at NASA's Jet Propulsion Laboratory, also utilises TAE. PDS provides archiving, user browsing and display capabilities for all NASA planetary missions.

User Assistant/Teaching Tool.

Better Image Software with Help On-Line for Programs (BISHOP) was developed by the Imperial College of London and uses TAE to create a user-friendly interface to the JPL Video Image Communication and Retrieval (VICAR) image processing system. The Imperial College has linked together other facility applications under TAE so that the novice user can use TAE menus and on-line help to get started easily on the system.

Because of its ease of use for novice users, several universities have built systems based on TAE for use by students with little or no computer background who are learning the applications provided in the system.

Defense Systems.

The Second Generation Comprehensive Helicopter Analysis System is being developed at the NASA/Ames Research Laboratory for the U.S. Army. It provides a computing environment for helicopter analysts and designers. TAE is being used as the foundation of the executive portion of this system.

The analysis support system for the Teal Ruby Experiment, from the Environmental Research Institute of Michigan, is using TAE to integrate and provide a human interface for a support system that is used in the storage, retrieval, processing and display of Teal Ruby Experiment Imagery. They are also using TAE for mission planning function interface.

Prototyping Tool.

At GSFC, TAE is being used to prototype different user interfaces for future real-time operator consoles at their mission control centers. Another prototyping effort at GSFC is using TAE to create user interfaces for Space Station, as a means of investigating and evaluating user interfaces for a spectrum of users.

3.6. Summary.

The software costs for an analysis system usually far outweigh the costs of the hardware itself. Reusable software, which is both portable and customisable, presents an alternative with a significant cost savings impact when it also provides an adequate set of services and a consistent, well designed user interface. TAE is a system builder's tool box that offers such an alternative.

A system designer may use TAE for a wide variety of applications because of its discipline independence. Users may become involved very early in a system design based on TAE by using the rapid prototyping capability. This gives the user a clear and early picture of what a system will look like and provides a good specification of the end product to the system builders.

Because of its open architecture, discipline independence and suitability to many environments, the TAE system will continue to grow. New capabilities are being added to meet the demands of the more than 200 user installations and to support new application areas.

TAE has become one of the principal tools of the Data Systems Technology Laboratory at the Goddard Space Flight Center, where it is used in rapid prototyping to support human factors research, and new applications in flight missions such as real-time operator consoles and Space Station applications. Broad usage and new demands continue to drive the growth of the TAE system.

Acknowledgements.

TAE is being developed by the NASA/Goddard Space Flight Center and by Century Computing, Inc. The work is sponsored by the NASA Office of Space Science and Applications and the Office of Space Operations

References.

Aho, A. V. and Johnson, S. C., 1974, LR parsing. ACM Computing Surveys 6:2, 99-124.
Anderson, J., 1979, HELP: Online Documentation System. Proceedings of the USE Spring Conference, pp. 215-235.
ANSI, 1979, OSCRL User Requirements. Rev. 7, ANSI X3H1/O5-SD.
ANSI, 1979, OSCRL Functional Requirements. Rev. 5, ANSI X3H1/06- SD.
ANSI, 1984, Operating System Command and Response Language (OSCRL) Language Specification (DRAFT), Rev. 18, ANSI X3H1/09- SD.
Beech, D., 1980, What is a command language? Command Language Directions, D. Beech, Ed. (New York: North Holland), pp. 7-27
Brown, P. J., 1977, Software Portability: An Advanced Course (Cambridge, MA.: Cambridge University Press)

Carlson, Patricia A., 1983, User-Programmer Dialogue: Guidelines for Designing Menu and Help Files for Interactive Computer Systems, NASA TM-84980, 51 pp.

CODASYL, 1982, CODASYL Common Opeerating Systems Command Language (COSCL) Journal of Development, Version 2.2. CODASYL Common Operating Systems Command Language Committee.

Cullingford, Richard E, Krueger, Myron W., Selfridge, Mallory and Marie A. Bienkowski, 1982, Automated Explanations as a Component of a Computer-aided Design System. IEEE Transactions on Systems, Man, and Cybernetics, SMC-12:2, 168-181.

Dalton, John T., Billingsley, James B., Quann, John J. and Bracken, Peter A., 1979, Interactive Color Map Displays of Domestic Information. ACM Computer Graphics 13:2, pp. 226-233

Dalton, John T., Jamros, Rita K., Helfer, Dorothy P. and Howell, David R., 1981, The Visible and Infrared Spin Scanning Radiometer (VISSR) Atmospheric Sounder (VAS) Ground Data Systems. Presented at the Society of Photo-Optical Instrumentation Engineers Technical Symposium (GSFC internal document).

Demers, Richard A, 1981, System Design for Usability. Communications of the ACM 24:8, 494-501.

desJardins, Mary and Petersen, Ralph, 1983, GEMPAK: An Interactive Display and Analysis System. Proceedings of the 9th Conference on Aerospace and Aeronautical Meterology, pp. 55-59. Engelberg, Norman and Shaw, Charles, 1984, Considerations of Command and Response Language Features for a Network of Heterogeneous Autonomous Computers," NASA TM 86089, 61 pp.

Fenchel, Robert S. and Estrin, Gerald, 1982, Self-describing Systems Using Integral Help. IEEE Transactions on Systems, Man, and Cybernetics, SMC-12:2, pp. 162-167.

Hall, Dennis, Scherrer, Deborah K. and Sventek, Joseph S., 1980, A Virtual Operating System, Communications of the ACM 23:9, 495- 502.

Howell, D. R., Owings, J., Szczur, M. R., Helfer, D. P., Lynch, D. M. and Cyprych, E. J., 1980, Conceptual Design for a Transportable Applications Executive, GSFC internal document.

Ling, Robert, 1980, General considerations on the design of an interactive system for data analysis. Communications of the ACM 23.3, 147-154.

Marcus, Aaron, 1982, Typographic design for interfaces of information systems. Proceedings of ACM Conference on Human Factors in Computer Systems, pp. 26-30.

Moran, Thomas P., 1981, An applied psychology of the user. ACM Computing Surveys 13:1, pp. 1-11

Perkins, Dorothy C., Szczur, Martha R., Owings, Jan and Jamros, Rita K., 1984, A Device Independent Interface for Image Display Software. Proceedings of the National Computer Graphics Association Conference, Vol. 2 pp. 392-401.

Relles, Nathan, Sondheimer, Norman K. and Ingargiola, Giorgio, 1981, A Unified Approach to Online Assistance. Proceedings of the National Computer Conference, pp. 383-388.

Relles, Nathan and Price, Lynne A., 1981, A User Interface for Online Assistance. Proceedings of the 5th International Conference on Software Engineering, pp. 400-408.

Roberts, Roger, 1970, HELP – A Question Answering System. Proceedings of the National Computer Conference, pp. 547-554. Scheifler, Robert W., Jim Gettys, "The X Window System," ACM Transactions on Graphics, Volume 5 No. 2, April 1987, pp. 79-109

Shneiderman, Benjamin, 1980, Software Psychology: Human Factors in Computer and Information Systems (Cambridge, MA.: Winthrop Publishers), 320 pp.

Snowberry, Kathleen, Parkinson, Stanley R. and Sission, Norwood, 1983, Computer display menus. Ergonomics 26:7, 699-712.

Szczur, Martha R., "TAE Plus: Providing a User Interface Development Environment," Proceedings of Defense and Government Computer Graphics Conference, September, 1987

Szczur, Martha R., Dorothy C. Perkins, David R. Howell and Karen L. Moe, "TAE Plus: An Integrated Design-to-Production Environment," Proceedings of the Ninth Annual National Computer Graphics Association Conference and Exposition, March 1988

Szczur, Martha R. and Philip Miller, "Transportable Applications Environment (TAE) Plus: Experiences in "Object"ively Modernizing a User Interface Environment," Proceedings of the Object-Oriented Programming: Systems, Languages and Applications Conference, September, 1988

4
A menu-based interface oriented to display processing of real-time satellite weather images

Kevin Tildsley and Chris England
Centre for Remote Sensing
Imperial College London

4.1. Introduction.

With the increasing use of expensive computer systems for remote sensing and image processing, many facilities are required to cater for users with little experience of computers. In many cases, these users are involved with relatively small image processing tasks, and it is neither cost nor time effective to train them to make full use of the system. A number of 'turnkey' systems have been developed to cater for such users, but these often suffer from problems of inadequate applications programs, and difficulties in systems extension to cater for developing requirements. Furthermore, such systems still require a significant learning effort before the user can fully interact with them.

This chapter presents details of an Interactive Menu Interface (IMI), which allows users to immediately undertake complex image processing tasks without any knowledge of either the host computer or the display devices. The IMI sits between the user and the applications software package, allowing the selection of appropriate programs and program parameters from a menu display. Selection is performed by means of a screen cursor, controlled by a pen and graphics tablet. Other means of menu interaction are supportable, but this mode was selected as it is easier and less tiring to use than either a mouse or trackerball.

The IMI is readily adaptable to most image processing systems, having been written with a high degree of system independence. Applications programs may be made compatible with the IMI by the inclusion of a standard interface routine, and may be linked in with the system by altering an initialisation file containing program parameters.

4.2. Menu Interaction—The User Viewpoint.

The Basic System.

To initialise the IMI, the pen is tapped on the graphics tablet. The system responds by writing a top-level menu, consisting of the program categories, at the bottom of the image display, in a vertical column on the left of the screen. Tapping one of the category names produces a list of the applications programs available within that category, displayed to the right of the category listing (Figure 4.1). The user may either change the category, by tapping another category word, or may select a program by tapping the relevant program word with the pen. On selecting a program, the menu is altered to display the program name to the right of the top-level menu, whilst three command words (**HELP**, **EXTRA** and **EXECUTE**) are written below.

The remainder of the menu area, consisting of a further four columns, now displays the program keywords, each associated with a parameter controlling program execution. Such keywords may be status words, indicating particular modes of operation, or they may be associated with alphanumeric input, e.g input/output file names or image size parameters. Since the selection of a particular program option, and hence keyword, may entail a requirement for differing inputs, the program display is based on a hierarchical tree structure, so that the selection of certain options may cause further menu words to be displayed. Tapping an option keyword a second time will deactivate that option, and cause any additional menu options to disappear. Additionally, if one of a number of conflicting options (e.g. **BLANK** and **PROTECT**) is selected, options incompatible with the latest selection may be either deactivated, or removed from the menu entirely.

To assist the user in keeping track of all the selected options, the keyword status is indicated by its changing colour on the display. In the present implementation, grey boxes are used to indicate inactive keywords, which are changed to blue when activated.

Menu Additions.

In many instances the programs have more options available than are generally required. To avoid the complexity of having large numbers of parameters displayed on the menu at any one time, many of them redundant for normal usage, an additional menu may be called up by the activation of the command word **EXTRA**. This will generate a further menu appearing at the top of the screen, containing the additional parameters for each program. Such a menu interacts in exactly the same fashion as the fundamental menu, except that it is uniquely a program menu, and is removed if a program or category change occurs before program execution. There are three command words, **QUIT**, **ENTER** and **RESTORE**, **ENTER** inputting any changes into the program, **RESTORE** removing any previous modifications and **QUIT** not changing anything. If such a menu has been called and basic parameters changed, the program menu, in the normal display mode, is altered by changing the background colour to indicate that changes have occurred in the program action.

An example of the use of such a menu may be provided by a program using a graphics box to define a window for program execution. The box would normally be constructed of lines of single pixel width. If the user wished to display the program operation, the ability to increase the line width to 2 pixels, to remove interlace flicker, would be an option selectable from the **EXTRA** menu.

Alphanumeric Input.

In contrast to logical and status options, which may be selected by merely tapping the word, numeric and character inputs require a further level of interaction, as discussed below:

Numerical Input.

When a keyword requiring numeric input is selected, the menu word is first tapped to activate the option, and the display is updated to show the default value of the parameter in cyan (as the value is active).If the default is desired, the display may be closed down, by a second tapping of the word, but if a change is required, the numeric value is tapped. This causes the display of a numeric keypad on the menu, showing the integers **0** to **9**, together with the maximum, minimum and default values of the parameter. A decimal point is included if the numeric type is real. A further three commands, **QUIT**, **ENTER** and **DELETE** are also displayed, whilst the menu word is surrounded by a white box to illustrate the word status. A box below the pad indicates the current value in the input buffer.

The user may now quit the procedure, wipe the buffer or enter the input buffer into the menu parameter by selecting one of the three commands. The value may also be updated by deleting the old value and entering a new number by tapping the appropriate value on the keypad. To minimise the possibility of error, the input buffer will not accept values outside the pre-defined range for the parameters, nor will it accept a null input. Additionally, the display will not allow a real number to be entered when an integer is required.

On entering the parameter the numeric pad disappears, leaving the display showing the new parameter value in place of the default. The display will retain this value, unless the menu word is again tapped, when the parameter will regain its default value, and the number will vanish from the display. Thus, the display will always show those input values which have been selected by the user, whilst defaulted variables are distinguished by having no displayed value.

File Input/Output.

On the majority of computer systems, a unique file specification will consist of a combination of the following;

a) the disk on which the file resides,
b) a main directory
c) a sub-directory
d) the name of the file
e) a specifier for the file type and
f) a version number, where several files of the same name may coexist.

All sections of such a specification being separated through a system defined syntax. On a VAX, for example, this might be;

    "DRD2:[FRED.PICTURES.METEOSAT]BRITAIN.IMA;4".

The ability to change all or part of such a file specification, without having to worry about the syntax is a major advantage for casual or novice users of the system. The procedure may be illustrated by considering the input/output of the above image file.

The user first selects the **LIBRARY** category from the top-level menu, whereupon the second level menu is displayed. This contains programs for the input and output of images, look-up tables, cursor shapes, windows, etc. from which, in this case, **IMAGE** would be selected. The display then presents the options for the program, in this case the general input/output commands **LOAD**, **SAVE**, **LIST**, **BROWSE** and **DELETE**, as well as parameters specific to image I/O, for example **SIZE** and **COLOUR**. Finally, an initial image name is displayed, for example, **FRED**.

FILE-NAME INPUT.

To change a file name, the user taps the current file name "FRED". This causes the menu to be replaced by an alphanumeric "keyboard", along with the keyboard commands **WIPE**, **QUIT**, **RUB-OUT** and **ENTER**, and a display of the input buffer, in a similar manner to the numeric input pad described above (see Figure 4.2). **WIPE** or **RUB-OUT** may be used to delete the whole buffer, or the last entry respectively, and the user may input the new name, in this case **BRITAIN** from the displayed keyboard. To return this name to the program, the command **ENTER** is selected, and the menu is replaced by the program menu with the new image name. If **QUIT** is selected, the menu will return to its previous display, without updating the parameters. As for numeric input, the program and display are set up so as to minimise errors. As the keyboard only displays valid characters for the name input, syntactical errors are impossible, whilst the buffer will not accept names longer than the permitted file name length (9 characters on a VAX).

LOCATION INPUT.

To specify the disk, directory and sub-directory input, a different approach has been taken, as the entry of long character strings (of indeterminate length) by keyboard is cumbersome and can lead to frequent mistakes, especially for a novice user. To enter these parameters, the IMI generates a list of available disks, and their directories, which may be selected by the user. To achieve this, in this illustration, the menu word IMAGE is tapped, to display a menu with the words:

```
DISK = CURRENT DISK
USER = CURRENT MAIN DIRECTORY
DIRECTORY = CURRENT SUB-DIRECTORY
FILE = CURRENT FILE
VERSION = LATEST
```

together with the commands **QUIT** and **ENTER**, with their usual functions.

Tapping the word **DISK** will generate a list of available disks, which may be selected by the user, the display reverting to the above listing once a disk name has been tapped. Tapping the **USER** word will produce a list of available directories on the current disk, which may be selected in the same way as above. Alternatively, tapping the current value will generate a keyboard for manual input of this parameter. A similar procedure is followed for the **DIRECTORY** keyword, except that no keyboard is returned, selection being entirely from the listing of the sub-directories contained within the **USER** main directory and the **DISK** disk. The version number is entered by the numeric keypad, in a similar manner to numeric input, although this will normally be the default, latest version. The actual image name may also be entered from this part of the program, following the procedure outlined above. A keypad option is presented, but for input files, users are encouraged to use the listing procedure in order to minimise the possibility of error. To prevent the user from selecting a valid directory/image on a given disk, and then changing the disk to one where these parameters are invalid, the menu will only allow selection from the top level downwards. Additionally, the user is not allowed to return to the main calling program unless the whole image specification is valid for input, or the directory specification is valid for output.

Another method of file selection uses the **LIST** and **BROWSE** commands within the **IMAGE** program menu. Selecting **LIST** will generate a list of all the images within the current disk and directory specifications, without the user having to fetch the input menu, whilst **BROWSE** will produce a display of images sub-sampled to fit 12 on the display screen. The user has only to tap the desired image, or image name, to load the image into the display. A **PAGE** (**UP/DOWN**) option allows the user to scan through large numbers of images.

4.3. Processes.

A powerful feature of the IMI lies in the ability of the user to define a process, here defined as a set of consecutive commands for repeated application, by simply running through the procedure once interactively. To create such a process, the user enters the **SYSTEM** category, and selects the **PROCESS** program, which displays a number of options, most importantly **START** and **STOP**. Having selected a name for the process, **START** is chosen and any further interaction between the user and the menu interface is recorded in a process file, which is terminated by a return to the **PROCESS** program, and hitting **STOP**. It should be noted that any activity permitted by the menus can be included within the process file, including the execution of further process files, which may be nested to any depth, subject of course to the restrictions imposed by the host computer system on available file space.

To retrieve a process for execution the process file is selected from within the **LIBRARY** menu and may be executed like any other menu program, repeating the procedures already laid down, but with differing inputs if required. Several modes of running processes exist, which may be selected at run time.

The simplest such mode is a pseudo-batch mode, where all functions selected in the process initialisation are automatically performed, bypassing the menu interface entirely, except for the input of file names for the second and later runs. To run totally independently of the menu, it is necessary to run the process in an 'input mode', where no application program execution takes place, the user being prompted solely for any input and output file names required by the process, together with the number of runs desired. An alternative, non-interactive mode, allowing the user to follow the procedure, performs the menu display operations during process execution, either to the program level, or deeper into the parameter selection. A further, semi-interactive mode of operation is available, where the user can run through the process, performing interactive processing within certain selected programs included within the process file. To enter this mode, the user must first run through the process once more, using a single step mode, where each program is executed and the user is prompted to see if interaction is desired within the program. This second step process definition also allows the user to edit the process file, by having options to include further programs, or exclude programs from the file, the user tapping **CONTINUE**, **ADD**, or **REMOVE** at each step to determine the next action.

An example of these modes of operation can be provided by considering cursor controlled brightness and contrast enhancement. In the 'batch' mode, the process will automatically apply the final enhancement used in the process definition. Menu display options will either produce the menu applicable to the program, or will follow the cursor operations of the original program run. In the interactive mode, the user will of course be able to select any enhancement, by adjusting the cursor position. Editing the process may allow the initial enhancement routine to be replaced by an enhancement within a selected brightness range, for example, or may exclude the enhancement completely. Such process definition allows users to build up their own library of interactive programs (within the limits of the applications programs) without having to make use of the host's control language, or having any knowledge of computing languages. Equally important is the fact that they require no knowledge of the editing facilities available on the host machine. The IMI is thus of great value for novice or infrequent users of a computer system. Another advantage of this approach to process definition lies in the method—since the process was defined by actually performing the operations, it is guaranteed to produce the desired result—a case of "what you see is what you get".

4.4. Menu Interaction—The Programmers View.

All the text and data required for the menu display are loaded from a set of data files into arrays when the user initialises the IMI. Each program has a unique identification number, based on the program category and its desired position in the menu display. This is used as an array index for obtaining program information. Each parameter is attached to a further identification number for array indexing. Five basic categories of parameter arrays are defined, namely refresh memory arrays, status flags, conditional arrays, general numeric parameters and file input/output parameters, as discussed below.

Fundamental Variables.

Status flags have the option of either on (-ve), or off (+ve), and control the menu display, indicating to the IMI whether the words should be displayed as either active or inactive. They are used internally at execution time to inform the programs of logical options, and whether certain parameters have been entered. For example, a program to draw a graphics box, to define a window, may either blank the graphics first, or not. This will be shown on the display by the word **BLANK** being either active or inactive. At execution the flag informs the applications program whether or not the blanking option has been selected. These parameters also control whether a word is displayed, the most significant bit (excluding the sign bit) being 1 for display and 0 for no display. Since the status of a given keyword may affect the status of other keywords within a program the status flag also contains an address within the conditional array . Thus if the value of the status flag is -16884, ($2^{14} + 500$),the word is displayed and active, deactivation requiring further action dictated by the contents of element 500 in the conditional array. If the word is deactivated, this becomes 16884, whilst if removed from the display the value is 500.This method was selected, rather than using a multi-dimensional array, in order to minimise the internal storage requirements of the program.

The conditional array contains a list of further parameters which may be either activated, deactivated, added or removed from the menu and from the parameter list transferred to the program at execution, depending on the value. The first element related to any given keyword status contains the number of elements within the array relating to that keyword. The following elements contain display information. Operation can of course be in either direction, depending on whether the 'calling' parameter has been either activated or not, the options being controlled by the values of the array elements. The array values operate in the same fashion as the status flags.

A refresh memory array contains a list of available and selected refresh memories within the display device and is attached to the parameter, both within the IMI internal arrays and on the display. On the display, the selected channels would be in cyan, whereas available channels would appear in red, with the selected channels being passed to the program at run time.

Numeric input parameters arrays contain five parameters. These are the maximum and minimum allowed values, the default value, the current value and whether the number is real or floating point. The data type descriptor controls the display of the numeric input pad, displaying a pad with a decimal point when required. The maxima and minima are checked on input, and values outside the allowed range are rejected. The status flag associated with the parameter informs the program whether the default or user input value is to be used.

File input and output parameters contain a list of file names, types, version numbers, directories and sub-directories. All array elements associated with a particular parameter are passed over to the applications programs, but the menu display is dependent on the status flags of differing program parameters, as discussed in section 4.2.

In addition to these parameter arrays, the menu text is loaded from disk files, along with housekeeping information including the number of characters in the words, the display position (line and sample) of each word, and the colour options for the menus.

Control Software.

The core software of the IMI comprises a loop containing routines for obtaining cursor coordinates, modifying the menus based upon these positions, obtaining the alphanumeric inputs and executing programs. The pen and tablet are repeatedly polled until the pen is tapped, when action is initiated depending on the cursor position. If tapped above the menu, the menu is removed allowing inspection of the image, whilst if tapped on the menu, a number of operations can take place, depending on the position of the cursor and program status.

If a new category name is tapped, the status flag of that category is set on, whilst all other category and program status flags are deactivated, resulting in the display of the applications programs within that category and the highlighting of the category word. Each of these changes are handled by the same code, and are purely dependent on the cursor position. If choosing a program from a category menu, the program status flag will be activated, causing the wiping of the category menu and its replacement with the program menu. At this stage, a different set of routines are invoked, in order to handle the file, channel and numeric display and input, together with the conditional array processing. As each menu word is written, the data type represented by each word is obtained by the routines, and current (or default) values are written to the display (depending on the word status flags) from the internal arrays.

If a selection is made at this level, the position flag is obtained, and action initiated depending on whether a command word (e.g. **HELP**, **EXECUTE**) or a program parameter has been activated. Since the routines know the number of characters in each menu word, it is simple to determine if either a word or an associated value (e.g. the "37" of the display **LINE**=37) has been selected, and either activate the keyword (by altering the status flag), or transfer control to the keypad routines. If a status flag has attached conditions, the conditional array is inspected, and action taken depending on the values.

When the **EXECUTE** command is issued from within the program menu, the arrays containing the status flags and parameter values are transferred to the program for execution. This may be be performed in a number of ways, either by spawning a sub-program, communicating with another process through a mail box, or by running the applications as sub-routines. The latter method has the great advantage of speed of execution, important for the use of such a system in an interactive manner, but suffers from a requirement to relink the IMI when new applications routines are included (with the exception of process files which may be constructed from the available programs). With the first two methods, the ability to change the menu display and actions by merely editing a data file allows the opportunity to add or remove programs, or to create totally different systems running from different initialisations. This can still be performed for the latter method, but loses some of its flexibility, owing to the need to relink.

4.5. Processes.

As discussed in Section 4.3, a process is a set of stored interactions between the IMI and the user, defined by actual program execution. The storage of the process is simple in principle, as all that is required is the storage of the $x - y$ coordinates of the cursor each time the cursor is polled. However, much of this information is redundant, as only a few of the cursor positions supply meaningful data to the calling programs (for example returning 5000 identical coordinate pairs merely because the user was thinking about which parameter to choose next is rather wasteful!). Therefore a certain amount of editing of these cursor positions is performed before storage. Thus, if a menu is being displayed, only those cursor positions where the user tapped the pen are saved in the process file. During program execution only coordinates different from the previous pair are retained. This may still lead to an excess of data storage within the program execution, but is essential to allow the user to see what actions were performed.

At the start of process definition, the disk space available to the user is checked to define an upper limit to the number of coordinates that may be stored, and the program will not allow process files to exceed this limit, by monitoring the available disk space as it is used up. Warnings are issued as the available space is depleted, allowing a user to exit from the procedure at a suitable point.

As a part of process definition it is possible to invoke a second process to run under the first. This is done by running the second process within the definition stage. The old process file data are copied over to the new file, rather than just including the second process name. This of course uses more disk space, but removes the possibility of process failure at a later date, in the event of the secondary process file having been deleted.

Editing the processes occurs in a similar manner. The process executes one program at a time, storing the coordinates in a temporary array, either writing them to a new file, or doing nothing, depending on the user selection. If a new program is required the temporary array is held while the new coordinates are written to the file at the completion of which they may be either deleted or copied into the new file, depending on the user keyword selection.

An important feature of the process files is the inclusion of 'prompt' points, created when the user calls either a listing, or a keypad routine. Numeric values and file names are included in the process file, but on execution the prompts cause the routines to drop back into an interactive mode, allowing users to change parameters if required. This prevents the process being locked to a particular set of input and output images, for example. Such prompt points are held in the file as numbers outside the range of the image display coordinates, to enable differentiation from valid cursor positions.

In 'batch' mode, it is of course not possible to make use of this method. Instead, the process is run initially ignoring all action except prompting the user for the number of runs required, and the relevant parameters at each prompt point, which are entered into a data file for later retrieval by the process at run time.

Execution of a process is simple, the IMI merely reads from the file, rather than polling the pen, ignoring selected levels of interaction depending on the operations mode, polling the tablet only at 'prompt' points, or at the end of each program if in the single step, editing mode.

4.6. Speed Of Execution.

In any interactive software, the speed of response to the user's commands is vital. Although the effect of other users on a time-sharing system is normally acceptable for terminal input, on a menu driven system it can be very frustrating to have to continually tap a parameter word for several seconds before the computer acknowledges the selection. This is in part a hardware limitation, as a normal pen and tablet do not retain information if a pen is tapped, implying that the pen must be polled at the time of the tap, whereas buttons on a mouse or trackerball box can keep this information until polled. However, the ease of use of a pen outweighs this disadvantage, and the software has been designed to overcome the problem. Thus, the polling of the pen takes place at a high priority, with a small wait between polling in order not to swamp the host system with polling commands, the rest of the program operation taking place at ordinary user priority.

To further increase the 'interactive speed', the display writing has been optimised to the nature of a raster system, writing complete lines across the display, rather than individual words at a time, by packing data into arrays before writing. Execution time is not, of course, any faster, but the system's response to menu interaction may be speeded up significantly by these methods.

4.7. Device Independence.

In order to make the IMI readily transferable to other computing systems, all of the routines have been written to be as device independent as possible. However, a number of features will need to be changed before they may be run on other computer systems, as discussed in the following sections. This machine independence does not of course apply to applications software, which makes use of specialised hardware and software within the machines, but relates to the ability to run the IMI for interactive control of those routines available on other systems.

Display Device Independence.

Hardware.

To transfer the IMI to another display device, a minimum hardware configuration is required. This minimum configuration is outlined below:

a) a number of image refresh memories (the number being dependent on the applications software),
b) an overlay graphics memory, viewed through a lookup table, and
c) a cursor control capable of returning the cursor coordinates and a further signal indicating that a selection has been made (for example a pen and graphics tablet which may return the information that the user has tapped the pen).

Software.

Three bottom level 'driver routines' are used by the system to control the display and display updates, which will require modification, or the inclusion of an intermediate level software interface before transferring the IMI to alternative devices. These are the routines that poll the cursor, read and write data to the displays, and assign colours to the menu graphics. No problems are foreseen with these modifications, as the data transferred is basic to most display devices available. The possible exception to this is the colouring routine, which may require further adaptation depending on available hardware.

Host Independence.

To minimise problems transferring the IMI to other host computers, the basic routines are written in standard FORTRAN 4. Those routines which require output from system services (such as file listing) are handled by intermediate level routines which transfer the appropriate data. Adaptation to other computers requires the re-writing of these routines. In the case of a directory listing, for example, this would require the writing of a sub-program to obtain the information from the host, and re-format it into the form required by the IMI. The number of such re-formatting routines is small, however, and should present no problems with most machines.

4.8. Applications Programs.

The design of the IMI has an impact on application program design. Firstly, it is important that the IMI philosophy is carried through to the programs. That is, it is pointless to use a user friendly system to run programs which are inflexible, prone to error and are not supplied with sufficient help information to assist the user with menu selection. Furthermore, where choices have to be made within a program, it is obviously desirable that a subset of the menu interactions should be employed to accept these choices rather than forcing the user to revert to keyboard input, which would defeat the whole object of the IMI. A number of such routines are available for inclusion in any applications program.

Secondly, the programs must be tailored to fulfil the functions available within the IMI. For example, running an applications program from a process file requires that the program be able to read from and write to the process file, where cursor coordinates are an integral part of the program control. Again, a standard cursor interface routine, performing these functions is available within the IMI. Similarly, the control of program options by transmitted logical and numeric parameters is essential to make full use of the menu capabilities.

Programs tailored to run under a number of systems are readily adaptable to this menu interface, in particular the Transportable Applications Executive (Helfer *et al.*, 1981), produced by NASA Goddard (see Chapter 3). This system has the added advantage that all of the menu definition functions and required parameters are available from the Parameter Definition File (PDF). All that is required to interface the programs to the IMI is a pre-processor for the PDF's to generate the initialisation files.

Acknowledgements.

We would like to thank the Centre for Remote Sensing, Imperial College, London, for their support in this work.

References.

Helfer, D.P *et. al.*, 1981, A Transportable Executive for Interactive Applications. *Proceedings of the Harvard Computer Graphics Conference.*

5
IAX—An Algebraic Image Processing Language for Research

Paul H. Jackson
IBM UK Laboratories Ltd
Hursley Park
Winchester, Hants

5.1. Introduction.

This chapter describes the IAX image processing language. IAX allows complex image processing operations to be specified in a simple and concise way. Two images A and B are added together to produce C in IAX by the statement C=A+B. The same expression is used to add two numbers (scalars), or two vectors. Scalars, vectors and images may be used in any combination in statements of this type. An image may be multiplied by a constant by the statement C=A*2. Expressions can be built up to arbitrary complexity. A more complex example, involving auto-correlation, is: C=IFT(ABS(FT(A))**2) where FT and IFT are the forward and inverse Fourier transform functions and **2 means raised to the power of 2.

IAX evolved over the period 1979–83 as part of the IAX image processing system and was designed and implemented by this author at the IBM UK Scientific Centre in Winchester (Jackson, 1983; Jackson, 1985a). It is an interactive interpretive system, originally developed for and used by image processing researchers.

The IAX system is a mainframe based software system which runs on IBM System 370 machines. The IAX language is at the centre of this system and is implemented as an interpreter.

Initially, IAX was developed as an internal tool. It was in actual use for most of its development period, on applications such as medical imaging. The system author was one of the major users.

The language evolved and changed often in the early days. Changes in compatibility were made when it was clear that the language was developing in the wrong way, but only while the user group was still small enough and there was not too great an investment in existing IAX programs.

Around 1982, the syntax of the language became quite stable and only upward compatible extensions have been made since then. Subsequently it has been taken up and used by IBM image processing researchers in many locations in several countries. More recently it has been installed in a number of universities and research establishments. It has been used in many applications, including: digital chest radiology (Cocklin, 1983b; ISMIII, 1982; Cocklin 1983c), light contour mapping (Gourlay, 1983a), coded aperture imaging (Gourlay, 1983b), texture analysis/pattern recognition, remote sensing (Muller, 1985), machine vision, image in publication, orthodontic radiology (Jackson, 1985b), CT brain scan analysis (Jackson, 1984) and speech signal analysis (Alderson, 1984).

This paper describes the present form of the IAX language rather than describing the development path and the forms of the language which were considered.

There are many useful features of the IAX implementation other than the IAX language. For example, it is very easy for users to write their own extensions to IAX, either as IAX procedures (macros) or as compiled code and this has allowed IAX to be extended in many ways. Although IAX is not based on any specific image display hardware, there are function packages for several of these. IAX has been used with Ramtek 9400 series, IBM 5080 display, Deanza, IBM 7350 image processing system, IBM 3270 display family and other devices supported by the GDDM graphical software package (GDDM), as well as various image hard copy devices. These and other aspects of IAX are described in references (Jackson, 1983; Jackson, 1985a and Cocklin, 1983a).

An image processing system is a form of 2 dimensional signal processing system. Although primarily a 2D system, IAX was designed also to be used as a 1D signal processing system.

IAX is implemented in the PL/1 language with some key routines in assembly language, in about 20,000 lines of code.

5.2. Languages Overview.

Because image processing is an application driven research area, the number of image processing systems for different specific applications is legion. Many of these systems are driven by a simple command language which can contain many commands. Although these are often termed "image processing languages" they are not what would normally be termed programming languages: they usually consist simply of a list of commands which exploits the hardware available. There is usually little possibility of programming an operation that the designer did not have in mind. An example of this kind is SUSIE (Batchelor, 1979). The majority of systems of this type (though not SUSIE Version 1) are programmed in FORTRAN.

Because of the poor string manipulation capabilities in this language, it is difficult to program anything other than a simple command interpreter.

A second type of image processing language is based on extensions to an existing language, or has a similar style to one of these. Examples of this type are PASCAL-PL (Uhr, 1981), which has parallel array processing extensions added to PASCAL, PIXAL (Onoe, 1981) and "L" (Radhakrishnan, 1981), both of which are extensions of ALGOL-60. Similar to this type are languages designed for special purpose hardware such as for the PICAP (Kruse, 1980) and GOP (Granlund, 1980) systems.

In these languages, images/array data types are often defined, although much of algorithm specification is carried out at the pixel level, with many loops and conditionals. Some high level array-oriented constructions are defined, but the structure of the underlying ALGOL or PASCAL can be inhibiting. For example, it is generally necessary to declare the sizes and types of all arrays, and these cannot conveniently be dynamically modified.

Perhaps a better starting point is APL (Iverson, 1962), (Gilman, 1974), since it is an array oriented language. The APLIAS system (Chanod, 1985) defines image processing extensions to APL, which are executed on an IBM 7350 image processing system. Currently APLIAS is only available for the IBM 7350. The array orientation of APL has many advantages for image processing, and it will be interesting to see how this approach develops. Possible drawbacks are the difficulty some people experience with the notation of APL, and the difficulty of adding compatible extensions to an already complex language.

There are many examples of image processing software systems. A good review of these is found in (Duff, 1981) and the many references therein.

The well known VICAR system provides another high level language (Seidman, 1979; Castleman, 1979; Jepson, 1976). VICAR is a mainframe based system, originating some 20 years ago. In VICAR, the effort has been put into providing a very large and comprehensive set of subroutines which perform image processing tasks. VICAR was built within an early OS type operating system, and inherits some of the complexities of this early type of operating system. Vicar is the image system containing the greatest manpower investment to date.

Implementation aspects of image processing languages can be very important, for example, whether the language is interpreted or compiled. Interactive interpretive systems often give enhanced usability in applications involving research. Another approach involves having no image manipulation language as such, where the system is entirely menu driven. An example of a highly interactive and well thought ought system of this type is PCIPS (Myers, 1985).

In other fields, more general programming languages have been designed which facilitate the manipulation of special kinds of data structures. The basic primitives of the language normally operate on such data structures.

In data base systems such as PRTV (Todd, 1976) the basic structure is a "relation" (which specifies a relationship between data in the database). In LISP (Winston, 1981), the programmer is concerned with list-structures and symbol manipulation, and in PROLOG (Clocksin, 1981), with logic structures.

In the IAX language, the basic structure is an image. Such structures are operated upon by special IAX image operators and functions.

The IAX language contains many features familiar in high level languages. It embodies the hierarchical operator and function notation for data manipulation as found in languages such as FORTRAN, ALGOL and PASCAL. It contains array processing primitives such as found in APL.

IAX brings together these approaches in an environment for image processing. Special image oriented features are defined and all operations are geared to the special requirements of image processing. In IAX, arrays are manipulated in a manner which is independent of their size and data type. Arrays may be created and destroyed dynamically with any data type and size, without using declarations.

5.3. Design Criteria.

IAX was designed to meet the following requirements:

1) to provide a simple, concise and yet powerful language for programming new image processing tasks, i.e. to facilitate image processing research and image algorithm prototyping.

2) to allow specific application oriented routines to be programmed easily.

3) to facilitate efficient internal implementation of primitive image processing functions.

4) to allow users easily to write their own extensions as compiled code, and to allow these to be automatically linked in to the system.

It is the intention of the remainder of this paper to demonstrate how the IAX language fulfills the first two of these criteria.

The third criterion is influenced by both the IAX language design and also by the internal implementation. As will be seen later, the IAX language is restricted to array data of 1 or 2 dimensions and to certain pre-defined data types. Both these factors facilitate an efficient implementation.

A fuller description of the way in which the IAX system meets the 3rd, and also the 4th criterion is given in the reference Jackson, 1985.

A feature of the design which became apparent as time progressed is that IAX is well suited to implementation for a special purpose array or image processor. This is because of the way in which computationally demanding operations appear in IAX as simple functions. An implementation of IAX using an FPS 164 array processor attached to an IBM 4341 mainframe is currently in progress.

5.4. IAX Language Concepts.

Programs in IAX consist of statements, normally separated by semicolons. A statement is one of five types and contains one or more expressions. The IAX statements are listed in Table 3.1.

Table 3.1 IAX statements.

Statement type	Example
Assignment	A=B+C
Command	TYPE(A,B,C)
Subscript assignment	A(0:10)=B
Pseudo-variable assignment	REAL(Z)=X
Control statement	DO I=1 TO 10;...; END;

Statements specify what is to be done with the results of expressions. An expression is the specification of a calculation to be performed. The result of the evaluation of an expression is a single data structure (an image for example).

Expressions consist of operators and/or functions applied to variables, literals, or subscripted expressions. Expressions may also contain further expressions. The components of expressions are defined as follows:

Operator.

A special symbol, such as + (addition operator), which specifies an operation to be performed on one or more expressions and which delivers a result

Function.

A named routine, such as SIN (sine function), which specifies an operation to be performed on one or more expressions and which delivers a result.

Variable.

A symbolic name for an item of data.

Literal.

The literal form of, or in other words the denotation for, an item of data.

Subscripted expression.

An expression which is followed by a special specification to define a subarray for array data, such as in X(55), which specifies the 55th column of image X or the 55th element of vector X. (Note that all array data is indexed from an origin of zero).

Examples of all these forms are to be found later in this Chapter.

Expressions specify calculations to be performed on data structures. IAX defines several types and formats of data structure. The following sections describe these data structures. Following this (in a bottom-up fashion), a more detailed description is given of the expressions in which these appear, followed by a description of the statements in which these expressions appear.

5.5. *Data Formats and Types of IAX Variables.*

IAX defines the following data formats:

Scalar	a single number
Vector	a 1-dimensional array
Image	a 2-dimensional array

These formats are implicitly inter-converted when necessary. For example, a row of an image is the same as a vector and an element of a vector the same as a scalar. Dyadic operations on any two of these formats involve the following:

1 conversion of the one with the less number of dimensions to the number of dimensions of the greater 2 conversion to the same number of elements 3 finally the operation is carried out on a element by element (i.e. pixel by pixel) basis to produce a result of the same size. (This is of course not necessarily the internal implementation).

There rules are known as the calculation rules.

IAX defines the following data types:

1 8 bit unsigned integer
2 16 bit signed integer
3 32 bit real (floating point)
4 64 bit complex (floating point)
5 character

The first four of these data types are normally termed the numeric types, distinguishing them from the character data type.

The number of bits per element is normally considered to be part of an implementation, not the language itself. However, image processing operations are computationally very taxing and efficient code generation is greatly facilitated if the data types are well defined in advance, as they are here.

As with the data formats, dyadic operations on different data types involve the conversion of what is termed the "narrower" data type to that of the "wider". The types given in the list above are from 1-narrowest to 5-widest. This is called the "widening rule".

In general, all operators and functions operate on all data types and formats.

5.6. Expressions.

As introduced above, IAX expressions may contain operators, function calls, special subscripted expressions, variables and literals. They may also contain other expressions. The first three of these specify operations to be performed on the variables and literals.

Variables.

An IAX variable can have any one of the formats and any one of the data types as described earlier under "Data Formats and Types of IAX Variables".

IAX variable names start with an alphabetic character followed by any number of alphanumeric characters (in our implementation there is an upper limit of 8 characters).

IAX variables are normally created by being assigned to, in an assignment statement. The type and format of a variable are not static attributes but may be changed at will. Consequently, there is no declaration statement. A variable can be changed from fixed to floating point format by the assignment A=FLOAT(A). A variable can even be increased in size by subscripting outside its current bounds, as in A=A(0:1023,0:1023). (in which it is supposed A is of size less than 1024×1024 before the assignment).

A variable may be initialised by a statement such as A=0(0:511,0:511), which defines a 512 by 512 integer image, with initial values of 0.

Literals.

A literal is a denotation for a data structure. The form of the literal determines the type and format of the data. The types are determined as follows:

    8 bit integer   no literal form

16 bit integer   normal integer (e.g. 55)
32 bit real   containing "." or "e" (e.g. 5.5, 1E6)
64 bit complex   of the form 4+5I
character   enclosed in quotes (e.g. 'ABCD')

The literal form for a scalar is a single number in one of the forms described above. (Note that a single character is a scalar). A character string is a vector of characters.

The literal form for a vector is a list of scalars (or expressions returning scalars) enclosed in brackets and separated by commas. For example, the following assignment defines a five element vector with elements 1, 2, 3, 4 and 5 respectively.

$$B=(1,2,3,4,5)$$

The literal form for an image is a list of vectors (or expressions returning vectors) enclosed in brackets and separated by commas. The following assignment defines a 3 by 3 image with values 1, 2, 3, 4, 5, 6, 7, 8 and 9.

$$B=((1,2,3),(4,5,6),(7,8,9))$$

Operators.

There are 18 IAX operators, which are described in Table 1. These are all array operators. All operate on every numeric data type and format (as described earlier under "Data Formats and Types of IAX Variables"). (There are some exceptions involving the character data type. For example, two character strings cannot be added together).

Operators return a single result and do not modify their operands. When applied to arrays, the operators return arrays. The Boolean operators (=, >, etc.) return Boolean arrays.

When dyadic operators are applied to any two from the set of scalars, vectors or images, the result always has the higher dimension. When operating on mixed types, the result always has the "wider" type.

Except for the concatenation operators, when operating on different sizes an operator's result is of the maximum size.

When converting from one type or format to another, the effect of an operator is identical to that which would be obtained if the programmer converted types explicitly using the built-in conversion functions, and changed size and/or number of dimensions explicitly to the maximum of the two operands using a subscripted expression (see "Subscripted expressions" later).

A few operators have their own special rules. These are the matrix multiply operator, which works according to the rules of matrix algebra, and the concatenation operators, which glue their two operands together to

Table 5.1. Operator Priority

Operator	Priority	Meaning
−	15	minus (monadic)
⌐	15	not (monadic)
**	10	power of
*	8	multiply
/	8	divide
!	8	matrix multiply
+	5	add (dyadic)
−	5	subtract (dyadic)
||	4	concatenate (horizontal)
!!	4	concatenate (vertical)
=	3	equal to
?	3	not equal to
≥	3	greater than or equal to
≤	3	less than or equal to
>	3	greater than
<	3	less than
&	1	and
|	1	or

form an array of greater size or number of dimensions (left-right in the case of || and top-bottom in the case of !!).

Functions.

All IAX functions are array functions. There are different classes of functions. About 80 are termed basic. These operate on all numeric data types and formats. There are many additional special purpose routines, some operating on only fixed point data for example. A summary of the functions is given later under "The IAX Functions and Commands".

IAX functions, for the most part, have only a small number of operands. They do not have large numbers of options because common options have been factored out and appear either as part of the language or as other functions. Functions may have a variable number of parameters: that is, the number of parameters may differ from one call to the next.

An example of the factoring out of common parameter options is subscripting. IAX functions which operate on images do not have "window" operands, i.e. four values which specify a rectangular sub-image. In IAX, *all* image operands may be subscripted as part of the IAX language. Hence, this facility is available with all functions (and of course operators).

Consider also a function which requires a non-complex variable. It may be used with a complex variable by applying the REAL, IMAG, ABS or

PHASE function to the complex variable. Functions therefore do not generally need to have options to describe what to do with complex operands.

In IAX, the function set is basic, simple and yet powerful because of the way the functions may be combined together and used with the facilities of the language.

Subscripted Expressions.

Subscripted expressions define sub-arrays of array data. They also define super-arrays i.e. larger arrays) and arrays with higher dimension (up to 2D).

Subscripted expressions are in one of the forms:

$$\langle expression\rangle \quad (\langle xsub\rangle,\langle ysub\rangle)$$
$$\langle expression\rangle \quad (\langle xsub\rangle)$$

where $\langle xsub\rangle$ and $\langle ysub\rangle$ are in either of the following forms

xsub/ysub	meaning
$p:q$	from p to q
$p:*$	from p to max
$*:q$	from min to q
p	row or column number p
$*$	all values in the dimension
$p :: \text{len}$	from p for len

p and q are subscripts and are themselves general expressions. The character : should be read as "to" and the characters :: as "for".

One dimensional data needs only the x-set of subscripts.

If an image, vector or scalar is subscripted outside its normal bounds, it is extended out to the required size by repetition of its edge elements (a scalar being considered a single element vector). Scalars may in this way be extended to form vectors. Both scalars and vectors may be extended to form images.

5.7. Statement Types.

Assignment.

An assignment specifies a variable to receive the result of an expression. In IAX, all of the attributes of a variable are determined by the result of the expression which is to be assigned to it. Variables do not have to be previously defined to be the target of an assignment. There is no declaration statement for specifying data type and format before an assignment. (Declaration statements are only used to define the scope of variables in subroutines. These are outside the scope of this paper).

If a variable is previously defined before being assigned to, the old attributes are overridden. The assignment statement may be used to allocate storage to a variable. When used in this way it is equivalent to a declaration. The following example defines a 256 by 256 pixel integer image, initialised to zero:

$$A=0(0:255,0:255)$$

Command.

Commands specify an operation to be performed on one or more expressions. The intermediate result of an expression is discarded after the command. Commands are like functions which deliver no result.

An example of a command is DISPLAY, which displays an image on the current default display device (varies according to hardware configuration) DISPLAY(A).

Subscript Assignment.

A subscript assignment places the result of an expression within a sub-array of an array variable. It may be used to set an element or group of elements of a vector, or an element, row, column or rectangular sub-array of an image.

Subscript assignments are subscripted expressions which appear on the left hand side of an assignment. The same forms of subscripts which appear in normal expressions are used, with one more additional form defined.

The full forms are:

$$\text{VAR}(X1 : X2, Y1 : Y2) = \langle expression \rangle$$
$$\text{VAR}(X1 :: \text{XLEN}, Y1 :: \text{YLEN}) = \langle expression \rangle$$

The meaning of the terms X1, Y2 etc are the same as defined above in the Section "Subscripted expressions". In addition, a new form is defined, in which X2 or Y2 may be specified as =. This form means "take the size given by the expression on the right of the assignment".

Psuedo·Variable Assignment.

In pseudo-variable assignment, a function applied to a variable is used as the target of an assignment. Because the function/variable combination takes the place of a variable, it is termed a pseudo-variable.

Pseudo-variable assignment is best illustrated by an example. In the following example, the variable Z is complex:

$$REAL(Z)=X; IMAG(Z)=Y;$$

Pseudo-variables can be more elaborate than this. For example, the AV pseudo-variable adds or subtracts a constant from an array so that the average value of the elements has a specified value, as in

$$AV(X)=25$$

The above statement is equivalent to the following statement in which the AV function appears:

$$X=X - AV(X)+25$$

The other IAX pseudo-variables are ABS, PHASE and SD (standard deviation). For each pseudo-variable there is a corresponding normal function.

Control Statements.

Control statements are part of constructs like DO loops and IF-THEN-ELSE conditionals.

The IAX DO loop is of a standard form:

$$DO \; \langle var \rangle = \langle start \rangle BY \; \langle step \rangle \; TO \; \langle end \rangle;$$
$$\langle statements \ldots \rangle;$$
$$END;$$

The conditional statement has the form

$$IF \; \langle Boolean \; expression \rangle THEN \langle statement \rangle$$
$$ELSE \langle statement \rangle$$

These are similar to (if in some cases subsets of) what is available in languages such as ALGOL and PL/1. There is no reason why this level of IAX should not be identical to one or other of these languages. For this reason there is no great value in defining these and other constructs like this more fully, as many precise definitions exist in these other conventional languages.

The implementation of IAX at IBM UK Scientific Centre is as an interpreter. This enables it to be used within the shell or EXEC language of the operating system. In the case of the VM operating system there are three of these EXEC languages, offering a variety of very flexible control constructions. We therefore use IAX in this way, rather than to develop the features of the IAX language into this area which is essentially independent of image processing.

5.8. Input and Output.

In IAX, input and output is carried out by functions. Hence, there is no I/O strictly defined in the language. The basic form of the input statement for placing an image into an IAX variable is:

$$A=GET('NAME', S1, S2, S3, S4\ldots);$$

The string 'NAME' is the name of the data file as known to the operating system. The parameters S1,S2 etc., are system dependent parameters. These may be used to specify the data type or format if the externally stored form of the image carries no specification of these.

The GET function may read data of any format or type.

The output operation is performed by means of a command called PUT, which is defined in a similar manner to GET.

5.9. Some Examples.

Operator Examples.

Here a simple example is described which illustrates the use of operators. The starting image is normally obtained in a manner such as:

$$A=GET('IM132');$$

where 'IM132' is the name of the image data file. The following statement halves all the pixels in A:
$$B=A/2;$$

Note that so far, the size, data type and data format of A are not known. Nor is it necessary to know this. A key feature of IAX is the way in which data is processed using statements which are independent of data size, type and format.

We can apply a Boolean expression to A as follows:

$$C=A<50 \; 3 \; A>100$$

C is an array consisting of 1's or 0's representing A less than 50 or greater than 100. If we wish to display C alongside A, we may ensure that these are in the same brightness range by multiplying C by the maximum value in A. This might be carried out as follows: C=C*MAXP(A). An example of a monadic operator is − (minus or negative). This can be used to negate a variable, as in the statement: D= −A;. The four variables A, B, C and D may be concatenated together as follows:

$$E=(A33B) !! (C33D);$$

The new variable E is thus twice as big as the others in both the X and Y dimensions. It contains A, B, C and D in the following positions: top left, top right, bottom left, bottom right. Figure 5.1 shows the result which is obtained if the above sequence of commands are applied to the 512×512 pixel test picture HH (Hursley House).

The annotation in the resultant picture was added for clarity, and does not form part of the resultant image. It is of course not essential to concatenate images together before displaying them. It has been done here in order not to describe any device dependent screen layout commands.

Function Examples.

Here some examples are described which illustrate the use of functions. Firstly some simple examples which illustrate the manipulation of vectors.

The VECTOR function generates a 1-D array (vector) with uniformly increasing values. Its three parameters specify start, step and end values in a similar manner to a DO loop.

$$X=VECTOR(1, 1, 10); \text{ TYPE } X;$$
$$\rightarrow 1 \quad 2 \quad 3 \quad 4 \quad 5 \quad 6 \quad 7 \quad 8 \quad 9 \quad 10$$

The symbol $\rightarrow$ is used here to indicate a response by the system. Remembering that all functions and operators may be applied to arrays, a variety of operations can easily be performed on X. For example, the statement:

$$\text{TYPE } 1.1**X$$
$$\rightarrow 1.1 \quad 1.21 \quad 1.331 \quad 1.4641 \quad 1.61051 \quad 1.77156 \ldots$$

generates a compound interest array, by raising 1.1 to the power of every element in the array X. A more complex example, which further illustrates the same principle, is:

$$Y=X**3 - 5*X**2 + 3*X - 5;$$

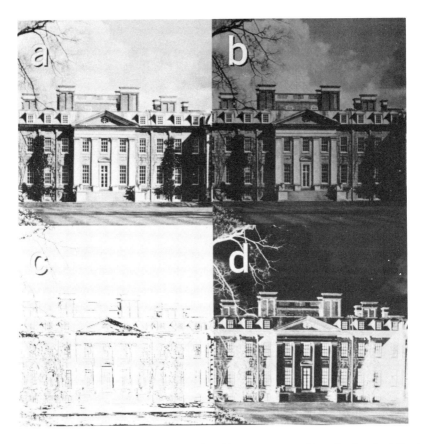

Figure 5.1. IAX Operators: The images B, C and D were generated from A using the following IAX statements

$$B = A/2; C = A < 503 A > 100; C = MAXP(A) * C; D = 255 - A;$$

Functions are applied to images in exactly the same way as to vectors. To obtain the absolute (magnitude) part of the Fourier transform of an image $IM1$, the following statement could be used:

IM2=ABS(FT(IM1));

The following example involves the processing of a satellite image. It is a NOAA 7 AVHRR, 11 micron, sea surface temperature image of the Alboran Gyre taken on 7th Oct 1982.

The purpose of this example is to show how some simple IAX functions can be used to form a mask which differentiates between the land and sea. This would be used as input to subsequent analysis of the sea temperature.

In this example the starting image will be called *A* and it is shown in Figure 5.2. (Note that *A* could just as easily be a vector and all the operations would then be 1-dimensional)

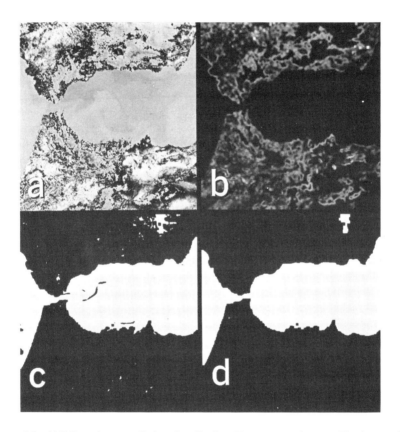

Figure 5.2. IAX Functions applied to Sea Surface Temperature Image: The images B, C and D were generated from the NOAA 7 AVHRR 11 micron sea surface temperature image A using the following IAX statements. The process provides an automatic filter for distinguishing between land and sea. B=LSD(A,7); C=255*(B≤1); D=LMAX(LMIN(C,7),7); D=LMIN(LMAX(D,7),7);

Firstly, the local standard deviation of A is obtained, as another image, as follows: B=LSD(A,7). The local standard deviation image contains the values of the standard deviation over a region (in this case 7 × 7 pixels wide) centred on the corresponding position in the original. Image B is also shown in Figure 5.2.

It is clear from B that the sea has a lower standard deviation of surface temperature than the land. Image C shows all values of B which are less than 1, determined as follows:

$$C=255*(B\leq 1);$$

C shows reasonable differentiation of the sea from the land, but there are some small misclassified regions. A useful technique for removing such "noise" is by use of the local minimum (LMIN) and local maximum (LMAX) operations (Nakagawa, 1978). LMIN replaces each pixel with the minimum neighbouring pixel value and therefore erodes white regions (makes them smaller). LMAX replaces each pixel with the maximum of its neighbours. If LMAX is applied after LMIN, it will restore any white regions which were not completely eroded away by the LMIN. Hence, to remove small white regions and leave larger white regions unchanged, the following IAX functions were applied to C

$$D=LMAX(LMIN(C,7),7);$$

The 7 refers to the size of the region over which the minimum or maximum is calculated. To remove small black regions and leave larger black regions unchanged, the LMIN and LMAX functions are applied in the reverse order:

$$D=LMIN(LMAX(D,7),7);$$

This final form of image D is shown in Figure 5.2.

Some further examples of IAX functions are shown in Figure 5.3.

Map Projection Example.

This example derives from work using IAX reported in (Muller, 1985). A in Figure 5.4 shows a METEOSAT II water vapour image.

In B the image has been map-projected to a rectangular coordinate system. Image C is derived from B and an image B0 taken some time earlier, as follows:

$$C=HE(B - B0);$$

The HE (histogram equalise) function has been used in this example to increase the contrast in the difference image. The difference image clearly shows the regions of greatest movement.

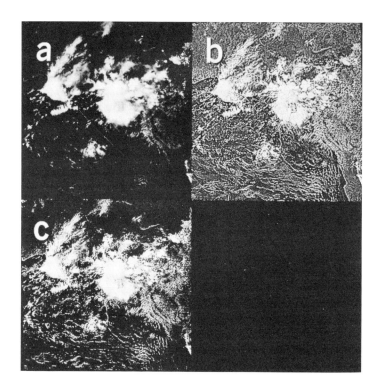

Figure 5.3. METEOSAT Image Enhancement: This shows a visible waveband Meteosat image A, and the results of applying to this the XSTA statistical differencing function (B) and the HP2 high pass filter function (C). This shows a visible waveband Meteosat image (a), and the results of applying to this the XSTA statistical differencing function (b) and the HP2 high pass filter function (c).

Example of Some Special Manipulation Functions.

In IAX, special functions such as THR, ROW, COL and SUBSCR are provided for data manipulation.

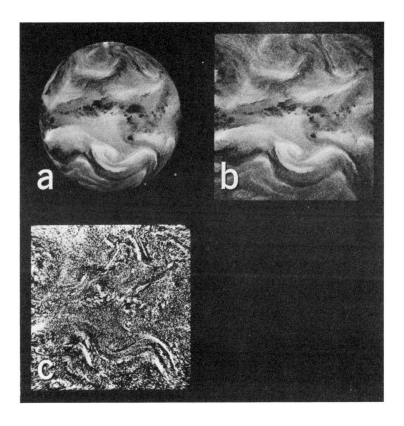

Figure 5.4. METEOSAT II Water Vapour Image Map Projection: The image A was subjected to a map projection process in order to give image B in a manner described in (Muller, 1985). The image C is a difference image between B and an image taken some time earlier.

THR.

The THR function, is a generalised threshold and merge function. It takes six parameters, as follows:

$$B=THR(A,P,Q,R,S,T);$$

The parameter A is the main parameter. The elements in A are individually compared with the elements in P and Q. If an element in A is within the range of the corresponding elements in P and Q i.e. ($\geq p$ or $\leq q$) the corresponding element in R is taken as the result. If it is less than this range, the corresponding element in S is taken, otherwise the corresponding element in T is taken.

With THR any of the parameters may be scalars, vectors or images. In the following example, all values which exceed 255 are reduced to 255, and all values which are less than 0 are increased to 0:

$$B=THR(A,0,255,A,0,255);$$

In the next example, M is a binary mask image containing 0's where the image A is to be selected, and 1's where A2 is to be selected:

$$B=THR(M,0,0,A,A,A2);$$

As there are no values in M which are less than 0 in this example, parameter 5, A, is not utilised.

The THR function THR(A,P,Q,R,S,T) can be expanded as an IAX expression as in the following statement:

$$B= (A \geq P 3 A \leq Q)*R + (A<P)*S + (A>Q)*T;$$

ROW, COL.

The ROW and COL functions apply 2-dimension oriented functions to, respectively, the rows and columns of a 2-d array.

To illustrate these, consider the SUM function, which delivers the sum of all the elements of an array. When applied to an image, SUM(A) returns a scalar. When used in conjunction with the ROW function, as in B=ROW(SUM,A); the result is a column vector containing the sums of all the rows of the image. The function COL operates in an analogous manner.

ROW and COL may be used with functions requiring two operands, or dyadic operators. In this case, ROW applies the function or operator successively between every pair of elements along the rows of the array. An example of this type of use is:

$$B=ROW(+,A);$$

This example is in fact equivalent to the previous example ROW(SUM,A).

ROW and COL help to extend the capabilities of IAX so that it can often be used in unexpected ways. For example, although there is no function for calculating a factorial, the expression ROW(*,VECTOR(1,1,N)) returns the factorial of N.

Perhaps more relevant to image processing is the use of ROW with functions such as FT. In the following example, B is the 2-d Fourier transform of A while C contains the Fourier transform of all the rows of A:

$$B=FT(A);C=ROW(FT,A);$$

5.10. Description of Functions, Commands & Pseudo-Variables.

The IAX Functions and Commands.

This section lists many of the IAX functions and commands. Most of these (the "basic" functions) operate on all data types and formats. Some others are restricted in certain ways (such as to integer data type only). A complete description of these is given in reference (Jackson, 1985).

Conversion functions.

conversion to 8/16 bit integer, real and complex, rounding and truncation, conversion to and from character string (10 functions).

Arithmetic functions.

SIN, COS, TAN, LOG, EXP, MOD, ABS, Matrix INVERSE, etc (about 20 functions).

General utility.

AV (average), SD (standard deviation), general statistics, histograms, histogram equalisation, MIN and MAX of two arrays, MIN and MAX element within an array, INDEX of entry (or entries) in array, mirror images, rotation, transpose, X and Y sizes of array, shifting, sorting, bit plane selection, vector generation.

Image scaling.

x and y dimension scaling by nearest neighbour, bilinear, interpolation, rubber sheet mapping, "anti-aliasing"; array element re-scaling to new min/max or new 5–95 percentile.

Filtering.

general convolution (any size mask, any data type or format), high pass filters, low pass, Butterworth filter, local min and max, median, unsharp mask filters, local mean and standard deviation, statistical differencing (Pratt, 1978), Laplace, Fourier domain filter generators, cross correlation.

Transforms.

Fourier, Hadamard, Haar transforms; Hamming, Hanning/Cosine bell windows.

Gradient operations.

DIFF (Image differentials d/dx and d/dy), DI (Direction code), SO (Sobel operator),RO (Roberts operator).

Complex.

REAL,IMAG, ABS, PHASE, CONJ (conjugate)

Look-up-table functions.

LUT for Hist-equalisation, contrast manipulations, etc, application of a LUT to an array, Colour image LUTs.

Manipulation functions.

ROW and COL (apply 2-D functions individually to the rows or columns of an array), THR (generalised threshold and merge), SUBSCR (generalised subscripting)

Half-toning and binary image.

Error diffusion, etc, (10 functions).

Boolean Functions.

ANY and EVERY convert Boolean arrays to scalars.

I/O.

GET and PUT.

The IAX Pseudo-Variables.

IAX pseudo-variables were described in an earlier section "Pseudo-variable assignment". This section briefly describes each pseudo-variable.

ABS.

sets the absolute part of a complex image, vector or scalar. The complex data (in real/imaginary format) is converted to abs/phase format, the absolute part is set to the required new value, and the data converted back to real/imaginary format.

AV.

adds or subtracts a constant to an image or vector to force its average to have a specified value. $AV(X)=B$ is equivalent to $X=X+B - AV(X)$.

IMAG.

allows the imaginary part of a complex image to be set while leaving the real part unchanged.

PHASE.

sets the phase part of a complex image, vector or scalar. The complex data (in real/imaginary format) is converted to abs/phase format, the phase part is set to the required new value, and the data converted back to real/imaginary format.

REAL.

allows the real part of a complex image to be set while leaving the imaginary part unchanged.

SD.

forces the standard deviation of an image or vector to a specified value. The average value is left unchanged. $SD(X)=B$ is equivalent to $X=(X - AV(X))*B/SD(X)+AV(X)$.

SUBSCR.

allows extended array subscripting, in which the elements of arrays are specified by a vector of subscripts.

References.

Alderson P. R., Kaye G., Lawrence S. G. C., Sinclair D., 1984, Speech Analyser based on the IBM Personal Computer, *Digital Signal Processing 84*, Capellini V and Constantinides A G(Eds), Elsevier Science Publishers B V (North-Holland).

Batchelor B. G., 1979, Interactive Image Analysis as a prototyping tool for industrial inspection, *Computers and Digital Techniques*, Vol 2(2), 61.

Castleman K. R., 1979, *Digital Image Processing*, Prentice-Hall Inc.

Chanod J. P., 1985, APLIAS, *Proceedings of Meeting on Image and Speech Understanding*, Paris, (April 1985) (available from IBM Paris Scientific Centre).

Clocksin W. F. and Mellish C. S., 1981, *Programming in Prolog*, Springer-Verlag, Heidelberg

Cocklin M. L, Gourlay A. R and Jackson P. H, 1983, The IAX Image Processing Language: RAMTEK 9400 Function Package, *IBM UKSC Report No 115*.

Cocklin M. L, Gourlay A. R, Jackson P. H, Kaye G, Kerr I and Lams P, 1983b, The Digital Processing of Chest Radiographs, *Image and Vision Computing*, Vol 1(2), 76.

Cocklin M, Gourlay A. R, Jackson P. H, Kaye G, Miessler M, Kerr I, Lams P, 1983c, An Image Processing System For Digital Chest X-Ray Images, *Medinfo-83*, van Bemmel, Ball and Wigertz (eds), IFIP-IMIA, North-Holland.

Duff M. J and Levaldi S, 1981, *Languages and Architectures for Image Processing*, Academic Press, London, (proceedings of a workshop held in Windsor 1979).

Graphical Data Display Manager: General Information, *IBM Publication*, Number SC33-0100.

Gilman, Leonard and Rose, 1974, Allen, *APL, An Interactive Approach*, Wiley, NY, (1974) 2nd Ed.

Gourlay A R, Denison D, Peacock A J and Morgan M D L, 1983a, Optical Computing—A Tool for the Four Dimensional Study of the Respiratory Function, *Medinfo-83*, van Bemmel, Ball and Wigertz (eds), IFIP-IMIA, North-Holland.

Gourlay A. R and Stephen J. B, 1983b, Geometric Coded Aperature Masks, *Applied Optics*, Vol 22, 4042.

Granlund G H, 1980, THE GOP, A Fast and Flexible Image Processor, *Proc 5th IAPR International Conf on Pattern Recognition*, Miami Beach.

Haralick R. M and Minden G., 1978, KANDIDATS, An Interactive Image Processing System, Computer Graphics and Image Processing, Vol 8, 1–15.

ISMIII, 1982, Five papers by the authors of Cocklin (1983b) presented at the *1st International Symposium on Medical Imaging and Image Interpretation*, Berlin, Oct., (published as one volume in IBM UKSC report no 110, 1982).

Iverson, Kenneth E, 1962, *A Programming Language*, Wiley, NY.

Jackson P. H, 1983, The IAX Image Processing Language, *IBM UKSC Report* No 113.

Jackson P. H and Lorber J, 1984, Brain and Ventricular Volume in Hydrocephalus, *Zeitschrift fur Kinderchirurgie*, Vol 39, Supplement II: 91–93.

Jackson P. H, 1985a, The IAX Image Processing System: Reference Manual, *IBM UKSC Report No 125*.

Jackson P H, Birnie D and Dickson G, 1985b, The Digital Processing of Cephalometric Radiographs, British Journal of Orthodontics, Vol 12, 122–132, (July 1985).

Jepson P. L, 1976, The Software/Hardware Interface for Interactive Image Processing at the Image Processing Laboratory of the Jet Propulsion Laboratory, *Proc Digital Equipment Computer Users Society*, Las Vegas, Nevada, 629.

Kruse B, Gudmundsson B and Antonsson D, 1980, FIP—The PICAP II filter processor, *Proc 5th IAPR International Conf on Pattern Recognition*, Miami Beach.

Muller, Jan-Peter, 1985, Applications of Pattern Recognition to Geophysical Fluid dynamic systems using remotely sensed data, *IBM UKSC Report*.

Myers H J and Bernstein R, 1985, Image Processing on the IBM Personal Computer, *Proceedings of the IEEE*, 1064–1070 (June 1985).

Nakagawa Y and Rosenfeld A, 1978, A Note on the use of local min and max operations in digital picture processing, *IEEE Trans SMC*, Vol SMC-8, 632.

Onoe, 1981, PIXAL: A high level language for image processing, in *Real-time Parallel computing*, M Onoe et. al. (Eds), Plenum Pub Co., New York (1981), pp 131–143.

Pratt W K, 1978, *Digital Image Processing*, Wiley, New York, 1978.

Radhakrishnan T, Barrera R, Guzman A, and Jinich A, 1981, A design of a High-level Language (L) for image processing, in *Duff (1981)*, pp25–40.

Seidman J. B and Smith A. Y, 1979, VICAR Image Processing System Guide to System Use, *JPL Publication*, No. 77–37, December 1979.

Todd S J P, 1976, The Peterlee Relational Test Vehicle—a System Overview, *IBM Systems Journal*, Vol 15(4), 285.

Uhr L, 1981, A language for parallel processing of arrays embedded in Pascal, in *Duff (1981)*, pp 53–88.

Winston P H and Horn B K P, 1981, *LISP*, Addison Wesley, Mass.

6
Microcomputers and Mass Storage Devices for Image Processing

D. C. Ferns* & N. P. Press,
Nigel Press Associates Ltd (DIAD Systems Ltd),
Edenbridge,
Kent TN8 6HS

6.1. Overview.

The characteristics of 'super' microcomputers are reviewed with reference to the requirements imposed by remote sensing users. The specifications of colour framestores, interfaces, RAM, data input peripherals, hard-copy output peripherals, and local area networks are evaluated with respect to processing LANDSAT and SPOT imagery. Recent developments in data storage and archiving media, including magtapes, floppy disks, tape cartridges and optical disks, are described. It is envisaged that the increasing performance and relatively lower costs of such microcomputer systems will be a major contributing factor to the more widespread operational use of remotely sensed imagery.

6.2. Introduction.

Following the success of personal computers in homes and small businesses, a new generation of more powerful microcomputers or 'supermicros' have begun to make a significant impact on technical and scientific markets, including that of image processing. The market for so-called mini-computers, such as the widely-used DEC PDP11. Microcomputers may be defined as desk-top-size, self-contained computer systems which are independently able to compute and display results.

---

* New address:
Logica Space & Defence Systems Ltd.,
Wyndham Court, Portsmouth Road, Cobham, Surrey.

The past decade has witnessed rapid technological development in computing hardware, producing continual reductions in the cost of systems. The high level microprocessors in super micros vastly outperform their forerunners of only two or three years ago, and include the ability of supporting concurrent multiple users. The strategy of established mainframe and mini-computer manufacturers has tended towards proprietary hardware, whereas only a limited number of semiconductor manufacturers have provided chips for use in supermicros. Of these the Motorola 68000 family and the Intel 86 series are the most widely used and popular. This has encouraged a higher degree of standardisation between several manufacturers. There are currently over 100 manufacturers of supermicros.

One indicator of advancing chip capabilities is wordsize—the number of bits (ones or zeros) a computer handles as a single entity. The larger the word size, the more efficient and powerful the microprocessor. The chips of the early 1970's had 4-bit words, while there are now several 32-bit chips in production. Although word size is a convenient measuring stick, there has in the past been a disagreement over how it should be determined. The criteria used include address size, the number of pins on the chip package, the size of the chip's data bus, and the size of its internal components, such as memory registers. The latter two are the most commonly used, and a standard nomenclature is now generally accepted. For instance, the Motorola 68000 chip in the stride (DIAD) supermicro is a '16/32 bit chip', having a 16 bit external data bus and 32 bit internal registers. Registers are single work memory locations used to temporarily hold data during programme execution. A 16/32 bit microprocessor with a typical clock speed of 8MHz is capable of 2 million operations per second.

A comparison of different microprocessor chips is provided in Table 6.1.

A parallel development has also occurred over the past decade, namely that almost all end users of satellite imagery now regard it as essential to process their image data digitally. When the first LANDSAT satellite was launched in 1972, almost all imagery was distributed in the form of photographic negatives or prints. However the quality of such products has proved to be grossly inferior to that of digitally processed images. Digital imagery allows users to work at larger scales and generate maps derived from processed imagery. It has therefore become necessary for all users of remotely sensed imagery to have access to a digital image processing system. At least 30 institutes, universities and companies in the UK now have access to such systems. The majority of these systems operate on mainframes or mini-computers, such as those manufactured by DEC (VAX, PDP11's), PRIME and Hewlett-Packard. A number of specialist companies offer system packages for processing remotely sensed data, based on the aforementioned computer (e.g. DIPIX (Canada), International Imaging Systems (USA), GEMS of Cambridge (UK), LogEtronics (USA), Vicom (USA)).

Table 6.1 Comparison of Microcomputer Chip Characteristics

	Chip	External data bus	Internal data bus
Zilog	Z-80	8	8
Intel	8088	8	16
Intel	8086	16	16
Motorola	68000	16	32
Zilog	Z-8001	16	16
Intel	IAPX-186	16	16
Intel	IAPX-286	16	16
Intel	IAPX-386	32	32
Motorola	68008	8	32
Motorola	68010	16	32
Motorola	68020	32	32

Figure 6.1. DIAD Systems 68K image processing system.

Industry has begun to offer similar turnkey systems on supermicrocomputers, offering comparable performance at significantly reduced costs.

The 16/32 bit microprocessors are available in a variety of computer packages, each with different specifications. The review below summarises typical hardware configurations as driven by the requirements of remote sensing users. Two commercially available systems will be used as examples for discussion. The first is a turnkey system offered by DIAD Systems, complete with menu-driven image processing software; and operating on a range of Motorola 68000 microcomputers manufactured by Stride Computers (Figure 6.1). The second is a networked micro-workstation system manufactured and distributed by Sun Microsystems, using Motorola 68010 microprocessors (Figure 6.2). Sun offer high resolution vector graphics software, which may be complemented by add-on hardware and software from International Imaging Systems (I2S), to produce an integrated system for satellite image processing.

The software packages required by the remote sensing community, for both image processing and geographical information systems, have been reviewed in the context of microcomputer developments by Bernstein *et al.*, 1984 and Ferns, 1984.

Figure 6.2. Sun Microsystems workstation terminal.

6.3. Microcomputer Hardware Configurations.

Colour Framestores.

The field of colour computer graphics has developed rapidly since systems first became commercially available at a realistic price in the mid-1970's. High resolution colour images can now be displayed and processed on microcomputers using a colour framestore, usually in the format of extra slot-in printed circuit boards or a separate unit box with an interface to the microcomputer.

Satellite data is received and archived in digital raster (scanned) format, and it is therefore appropriate to use the raster scan technology utilised by many such framestores. The quality of the image display is largely determined by two factors, namely the number of bit planes or colours which may be displayed, and the size of image which can be displayed on the monitor. The majority of micro-based colour monitors accept red, green and blue (RGB) video inputs, and the number of bit planes in each of three channels relates directly to the potential colour display quality (Table 6.2).

Table 6.2 The number of bit planes in a colour framestore related to the maximum number of display colours.

Number (n)	Max Colours
8	256
12	4096
16	65536
24	16.7 million

LANDSAT data is received from the satellite with 6 bits/pixel, but the widely available computer compatible tapes (CCTs) are formatted and archived with 8 bits/pixel. LANDSAT MSS imagery has four 8 bit bands, each 3240 × 2340 pixels, and the higher resolution LANDSAT TM data has six 8 bit bands, each 6468 × 5375 pixels. Almost all interactive satellite image processing systems allow the user to display any three chosen bands through the red, green and blue channels, creating a colour composite image. It is only on a 24 bit plane system, or its hardware equivalent, that the unaltered raw LANDSAT data can be displayed with, 8 planes per band. Using systems with less than 24 bit planes requires some form of data compression, usually by ignoring the least significant bits of the raw data (rightmost of the binary number). However, the radiance levels of the LANDSAT data are rarely distributed over the 255 grey levels (2^8 bits), and programs to selectively reduce the data to 5 or 6 bits/pixel with no loss of information may be implemented for certain data sources.

The human eye is only able to discern limited numbers of distinct colours, and satellite data may therefore be reduced down to four or five bits/channel without noticeably affecting the quality of a three band colour display image. Thus a 12 bit plane framestore may have apparently similar colour display qualities to a 24 bit framestore, but it is possible that the minor differences could be those which are important to the interpreter.

Graphic overlay planes may be included within the confines of the framestore, with monochrome overlay each occupying single bit planes and colour overlays typically built up to 8 bit planes (256 colours).

The size of image display on medium to high resolution monitors is typically 256^2, 512^2, 1024^2 pixels, displayed on either 13" of 20" display monitors. 1024×1024 pixels displays have only recently become available at reasonable cost, and 512^2 pixel displays remain the dominant standard, even on mainframe and mini image processing systems. The availability of 1024^2 pixel displays is an important advance for many remote sensing applications. LANDSAT TM and multispectral SPOT-HRV imagery have a pixel ground resolution of 30 and 20m^2 respectively, and thus a 1024^2 display is required to view a similar spatial area to a 512^2 LANDSAT MSS scene (Table 6.3).

Table 6.3 Sub-sample intervals to display full LANDSAT MSS and TM images on colour display devices (512 × 512 and 1024 × 1024).

	MSS		TM		SPOT	
	Column	Row	Column	Row	Column	Row
Number of pixels per full scene column/row	3240	2340	6468	5375	6000	6000
Sub-sample interval to display on 512 × 512	7	5	13	11	12	12
Sub-sample interval to display on 1024 × 1024	4	3	7	6	6	6

The majority of image processing systems for satellite image processing have 512^2 displays. Unless these displays (and framestores) are upgraded to 1024^2, then the improved spatial resolution of the LANDSAT TM and SPOT-HRV satellite sensors will be traded against the poorer ability of the satellite image to give an overview. A high resolution image which is sub-sampled on the display device is a less accurate picture of a large area than would be a full pixel display of the same spatial area as seen on a satellite image of lower spatial resolution. In addition, the smaller the instantaneous field of view (IFOV) of the sensor (Townshend, 1981), the less representative each pixel becomes of its general surrounding area; and as a consequence a high resolution image which is sub-sampled to fit the display device will make considerably less sense and will be aesthetically less valuable than a low resolution image.

It is of course possible to reduce this problem by using spatial filtering (box averaging) software as a data compression technique, which would allow presentable image screen displays of, say, a full LANDSAT TM image. The problem of sub-sampling screen displays and framestore resolution is quantified by examining the sub-sample intervals in Table 6.3.

Interfaces.

'Interface' is a general term to describe a connecting link either between two systems, or between a system and some peripheral device. Interfaces usually require both hardware and software, but the hardware interface parts usually conform to widely accepted international standards for microcomputers such as RS232 (serial interface), "Centronics" (parallel interface) and the more complex and versatile IEEE-488 parallel interface. RS232 intefaces are provided with many peripherals devices used in image processing systems, such as terminals, trackerballs, mice, graphics tablets and links to host computers which can for example, be utilised to load satellite imagery onto the microsystem. The disadvantage of serial line interfaces however is the speed at which data is transmitted and received, known as the "baud rate". The speed in bauds is the number of indiscrete signal events/s. The maximum transmission speeds down RS232 lines are commonly about 20 Kbauds, but when providing a link to a host computer, are more normally used at 9.6 Kbauds. Whilst a serial interface transmits data one bit at a time using a 3 wire system, a parallel interface such as the Centronics is able to send eight bits simultaneously using a 10 wire cable. Centronics interfaces are commonly used to link to printers and can of course transmit data more rapidly than a RS232 serial line. More complex parallel interfaces are able to transmit and receive data of rates of up to 1.5 Mbaud.

RAM—Random Access Memory.

Random access memory (RAM) is the computer memory chip, or chips, which may be read from or written to in a random access manner, implying that access time to information from any location is the same. The image processing software is usually stored on disk and selected programs are loaded into RAM to allow the microprocessor faster access times to run program instructions. Some sytems also use RAM to temporarily store processed image data, termed "look up table". The RAM in microcompputers is frequently expandable by adding more memory chips. The complexity and length of high resolution image processing software suites requires a minimum of about 150 KBytes RAM to configure a functional high resolution image processing system.

Although the architecture or structure of different systems is diverse, a larger RAM capacity generally indicates greater versatility and possibly room for an additional suite of programs, such as satellite based geographical information system (GIS) software.

Peripherals.

Interactive image processing systems require at least one item of equipment from each of two groups of peripherals in order to be fully effective in remote sensing applications, namely information digitisers and hard copy output devices.

Data Input Peripherals.

Analysis of satellite imagery cannot provide accurate maps or complete solutions to environmental problems, unless supplementary information from other sources is also used. Many other sources of information may be regarded as spatial databases, and it is becoming increasingly important to integrate other data sets on the image analysis computer.

Firstly, systems require a means to input non-satellite data in a form which may be directly related to the satellite imagery. Typically this may be a digitising tablet which enables the user to manually digitise co-ordinates and vectors relating to map data. In this way the user may produce digitial overlays of specific map data, and initiate an embryonic Geographical Information System (GIS) on the satellite image base. Some image processing systems allow the user to digitise ground control points (gcps) directly from a map, and the image may be geometrically warped to fit the required map projection. The size of digitising tablets varies from A4 to approximately 4ft × 6ft, and as a general rule gcps digitised from a larger map scale will provide a more accurate correction procedure than those digitised from smaller scale maps on a smaller digitising tablet.

Manual digitising of data has several problems. It relies on human dexterity and is thus subject to human errors, and careful digitising is extremely time-consuming and labour-intensive. Furthermore, manually digitised data is only easily stored in a vector format, and because the majority of image processing systems work with raster images, it is necessary to convert the vector data to raster data before it can be stored in a compatible format.

This problem may be overcome by the use of a video digitiser , which uses the inherent raster of a television camera system to sample an image or

Table 6.4 Hard copy devices of reproducing colour satellite imagery from a display monitor. These are listed in descending order of quality and cost, as perceived by the author.

Colour filmwriter/plotter	*****	*****
Black and white filmwriter/plotter	*****	****
Matrix camera systems	***	***
Polaroid matrix camera	***	***
Ink-jet printer	**	***
Colour dot matrix printers	*	**
Camera (35 mm) screen photos	**	*

map. Low cost high resolution video digitisers are now available, typically including the following components:
- black and white television camera
- frame grabber
- vertical mounting camera stand
- computer interface

The image or map viewed by the camera is converted by the frame grabber from an analogue to a digital signal in near real time, with the resolution controlled by the capabilities of the colour framestore device. The current DIAD digitiser, for example, produces a 512×512 (or up to 640×576) raster image of 8 bits per pixel. A fourfold increase in accuracy can be expected when frame grabbers are developed for 1024×1024 bit systems.

Video digitisers are certain to have an increasing role in remote sensing applications facilitating simple digitising of aerial photographs, space photography and a variety of map products. Software for precise geometric correction and registration of digitised aerial photographs to satellite imagery will soon be available at acceptable costs, and video digitisers are likely to promote geographical information systems which include combinations of satellite imagery, aerial photography, and map data.

Hard-copy Output Peripherals.

The human skills of image interpretation which follow digital processing are best carried out using film or print copy, rather than utilising time on the computer system and thus prevent it from being used for its processing functions.

This second group of peripherals complementary to the image processing system includes a wide variety of 'hard-copy' output devices. These are listed in subjective order of image quality and cost in Table 6.4.

In the production of large format colour satellite prints, the quality of expensive filmwriter/plotter machines cannot be equalled in terms of cartographic accuracy, potential image sheet size and colour control. Unfortunately new technology has not made significant impact on the cost of filmwriters, which continue to be essential for all high quality image output. As a result only three commercial UK companies own filmwriters for remote sensing applications. Consultants and commercial organisations use filmwriters almost to the exclusion of other devices.

Polaroid colour matrix cameras, capable of rapid production of 9 inch square colour prints and colour overhead projection slides, may also be used for medium quality satellite image reproduction.

The cost of colour dot matrix (or serial impact) printers with high resolution graphics capability has decreased to about £500 for a relatively slow, 32 colour-shade printer. A wider spectrum of colours, finer dots and faster print speeds are features of more expensive models.

Ink-jet plotters are now having a considerable impact on the middle range of output devices, with printing resolutions of 140 dots per inch now available on machines printing a colour image of 8" × 11" in under two minutes. Whereas a dot matrix printer uses a needle or 'hammer' to produce its dot by impact, the ink jet system forms its dots by droplets of ink. This is done by supplying voltage pulses across piezo electric crystals which are formed around each nozzle. When a crystal receives a pulse its shape changes for long enough to increase the pressure in the nozzle. This forces a droplet of ink forward and out of the nozzle onto the paper.

These new hard copy devices will be ideal for reproducing educational imagery. The limited paper size acceptable to both dot matrix and ink-jet printers restricts the maximum size of image which may be printed, and thus the versatility of these printers in plotting satellite imagery is somewhat limited. Filmwriters are rarely connected to micro image processing systems for purely economic reasons, but all other types of output device mentioned could easily be integrated to a micro-based system.

Camera photographs of the monitor screen are probably the most widely used method of obtaining hard-copy, frequently in the form of colour transparencies. The major drawback of such output is geometric distortion, from the curvature of the monitor screen, camera lens and camera position. Distortion is considerably reduced by utilising a matrix camera unit which receives the digital signal from the RGB video leads, and a camera within the closed unit records the image through an optical system which considerably reduces the effects of distortion.

Computer networks.

A network is a line of communication between physically scattered terminals and computers, expressed in both software and hardware. The speed and effectiveness of network communications varies enormously. The most efficient networks for high resolution image processing require a combination of micro/minicomputer workstations, typically 16 or 32 bit, which are interconnected, either to a mainframe or to each other. Networks allow data to be passed between microcomputer nodes in self-contained messages or "packets", and certain systems allow the combined data and devices of the full network ring to be accessed from any one terminal; in other words the equivalent of a large mainframe computer can be physically distributed amongst the workstations.

The Sun Microsystems 32-bit graphics microworkstations utilise the Ethernet local area workstation (LAN).

Ethernet is now a widely accepted network with interfaces to a large number of computers from different manufacturers. Data transmission speeds of 10–12 Mbps are usually attained between nodes, typically to exchange information, access peripherals and gateways.

It is not difficult to see the value of this type of system for remote sensing and GIS applications. The addition of further minis or micros to a network adds to the mutually accessible storage space and data, whereas additional microcomputer terminals linked to a host mainframe computer may deplete precious mainframe CPU time and disk space, thus reducing the processing speed and the mean data storage per terminal.

6.4. Storage and Archiving Media.

Raw Data.

Image data from the LANDSAT satellites are supplied and purchased almost exclusively on 9-track, ½" magnetic tape, recorded at 1600bpi (bits/in), and occupying 10½" reels containing 2400ft of tape. LANDSAT 4 and 5 LANDSAT TM data is however increasingly available on higher density 6250 bpi tapes. There are several ground receiving stations for this data, and each station produces similar formatted computer compatable tapes (CCTs), which are purchased by the users.

The storage capacities required by image data from LANDSAT MSS and LANDSAT TM data, and SPOT-HRV data are summarised in table 6.5, accounting for full scenes and common subscene extracts.

Should the system supplier or end user require to store data on other media, the costs of data transcription from CCT need to be considered. Thus there is a demand for micro image processing systems which communicate

Table 6.5 Approximate storage capacity (MBytes-MB) required by full LANDSAT and SPOT scenes, and common extracts.

	MSS	TM	SPOT
Full Scene			
Storage required by a single band	8 MB	35 MB	9 MB + 36 MB
Number of bands	4	7	3+1
Storage required by all bands	32 MB	245 MB	63 MB
Minimum number of 1600 bpi 10½" tapes	1	4	2
512 × 512 Sub-scene			
Storage required by a single band	0.25 MB	0.25 MB	0.25 MB
Storage required by all bands	1.00 MB	1.50 MB	0.45 MB
Fraction of total image area	$1/_{29}$th	$1/_{132}$nd	$1/_{137}$th
Percentage of total image area	3.48%	0.75%	0.73%
1024 × 1024 Sub-scene			
Storage required by a single band	1.00 MB	1.00 MB	1.00 MB
Storage required by all bands	4.00 MB	7.0 MB	1.75 MB
Fraction of total image area	$1/_{7}$th	$1/_{33}$rd	$1/_{34}$th
Percentage of total image area	14.28%	3.03%	2.91%

with a reel-to-reel tape drive, enabling CCTs to be loaded and processed.

System Storage.

Integral Hard/Winchester Disks.

The microcomputer must be able at least to store on the disk the minimum image data which the user required to display and process, which may typically be a 4 band LANDSAT MSS sub-scene, 512^2 pixels, requiring approximately 1.0 Mbyte of storage space.

The larger the storage capacity of a Winchester disk, the more convenient the system is to use, as more images may be stored without recourse to loading new images from the media archive.

New higher density storage methods on traditional Winchester disks, have allowed the packaged size to be decreased while storage capacity can be further increased. This trend has already taken place, a reduction in the size (width) of Winchesters from 14" to 8", then to 5¼", and now to 3½". 5¼" media have proven particularly successful with the computer industry and end users, and the widest range of manufactured products currently conforms to the 5¼" floppy format.

5¼" Winchester disks are available with capacities ranging from 5 to 120 Mbytes, and are usually integrated into the microcomputer system package. Some manufacturers have bus backplanes (e.g. SASSI bus on Stride 460) which allow up to 4 disks, in this case, each of 110 Mbytes, to be configured within the supermicro system. This 440 Mbytes of storage on a 68000 supermicro is a configuration competitive with many more expensive minicomputers, and is outside the realm of performance which many expect from micro.

Micro-Mainframe Connections.

As discussed earlier, interfaces, can be used to connect microcomputer workstations to an existing mainframe or minicomputer, thus providing access to the larger hard disk storage facilities on the mainframe. The same interface may be used to load raw data to the microcomputer from a mainframe tape drive. Low cost RS232 or IEEE 488 interface standards might be typical for such links.

Network Storage Access.

As discussed earlier, microcomputers in local area ring network configurations can access hard disk facilities at other network nodes. Typically, one node would act as a 'fileserver', with a larger hard disk and image loading facilities. The 'fileserver' could be a supermicro node, or a mini or mainframe computer with larger resources. Any node can also access the hard disk of any other node by requesting a file search.

System Input and Archive Media.

Tape Drives.

As previously stated, the raw satellite data is recorded and disseminated almost exclusively on 1600/6250 bpi 9-track, 10½" reel tapes. Few general purpose microcomputers support tape drive controllers, but there are now a small number of companies providing this service, even to 8/16 bit microcomputers such as the IBM PC. Tape loading software should allow users to spatially select small areas of imagery for extraction and loading to the Winchester disk. An RS232/IEEE488 link to an existing mainframe computer with disk and tape drive may also be initiated as the data loading mechanism.

Floppy disks.

Despite their limited capacity and less than perfect reliability, floppy disks are the dominant archiving media for microcomputers, not least because relatively low-cost floppy disk drives are a standard, integral component of almost all business microcomputer systems. Floppy disks are available in 8" format, 5¼" format and 3" or 3½" micro-floppy format. Floppy disks are formatted in the diskdrive to divide the disk surface into a number of sectors. There is considerable variation in the storage capacity of floppy disks depending on the formatting induced by the particular disk drive, with disk capacities ranging from 100 to 1500 KBytes (1.5 MBytes). Image data would normally be stored on double-sided, double-density floppies, to keep the number of disks to a minimum. Disk formatting also tends to preclude the compatibility between data stored on floppies by different systems, and thus floppies containing satellite image data are available only from the appropriate system manufacturer or agency. Table 6.5 illustrates the characteristics of image data floppy disks transcribed by NASA (EROS) and NPA Ltd.

Systems which are configured to input data only from floppy disks may be adequate for training in higher education. The limitation, however, is the inability to display and process a full LANDSAT image.

Tape Cartridges.

Although not yet widely used as a data archiving media, 5¼" tape cartridge drives with capacities of 15–60 MBytes have established a major market niche for winchester disk back-up.

Standard ¼" cartridges offer a unique combination of low media costs, high capacities, adequate transfer rates, compact size and virtually foolproof operation, and they are available in a variety of tape cartridges. Industry standard 4 inch by 6 inch cartridges with 450 feet of tape can store up

Table 6.6 Format specifications of LANDSAT data transcription services onto floppy disks.

	NASA (EROS)*	NPA Ltd
Floppy disk characteristics	8" Single sided Single density 256 KBytes	5¼" Single sided Double density 640 KBytes
Market capture	Spectral Data RIPS	DIAD 68K
Size of LANDSAT extract	240 × 256	512 × 512
Number of floppies per 4 band scene	1	2
Cost of 4 MSS band extract	$100	£70
Cost of additional sub-scenes from same CCT ordered at same time	$30	£30
Availability	EROS Archive MSS Post-1979 only	NPA open file Archive Any tape held, or any tape costed to client or owned client

to 45 MBytes of data in a streaming format. With the 600 foot cartridge, capacities of over 60 MBytes are possible.

Optical Disks.

A limited number of companies is currently involved in the development of optical disk technology. The majority of these commercial organisations, are small companies, and their product development has been notably slower than anticipated throughout the industry. The 'euphoria' of the publicity has been marred by a number of factors, including:
- shortage of disk supplies for optical drives
- lack of standard computer interfaces
- lack of recognised standards for disk size, disk capacity, and disk formats
- failure of companies to deliver 'available' products on time
- failure of companies to keep to promises of launching new,

more useful optical products (ie poor funding of product research due to lower than anticipated sales?). The single optical disk drives which are

---

* Specifications extracted from LANDSAT DATA Users Notes, No. 29, December 1983

currently available are of approximately 800 MBytes to 3 GBytes capacity. All currently available disks are read only, the cannot be interactively edited or overwritten. By the same token, optical disks are not eraseable. Edits to stored material must be done by amendment files. Products are being released without interfaces to computers, and thus original equipment manufacturers (OEMs) and scientific institutes are developing their own for resale.

Although most interfaces for optical disk drives are being developed to mainframe computers, there are interfaces already under development to super micro-systems. The Space Telescope Science Institute (Baltimore, MD) is, for example, developing a SCUSCI interface between a Sun workstation network and a Shugart Optimen optical disk drive.

Read-write disk technology is in the early stage of research, and while there appears to be a high degree of confidence in this technology, there is as yet no operational prototype.

Optical disk carousels or juke boxes provide a mechanical robotic arm which can pick up any disk from a storage rack, typically containing 50 to 150 disks, and load it into the single disk drive. This technology is currently in the prototype stage with a handful of companies and, on the previous record of the industry, it is anticipated that it will be 4 to 5 years before these products are commercially available as production units.

The use of optical disks may well replace magnetic tape as the storage and dissemination medium, but not before the industry sets standard formats which are widely acceptable to end users.

Indeed, for any alternative storage media to succeed in the remote sensing community, it must be launched by the data distributors, such as NOAA/EROS, ESA and SPOT IMAGE. Without distributor backing the cost of data on alternative media must be doubly borne by the end user.

6.5. Conclusions.

Advances in microcomputer colour image processing workstations and storage technology have not yet been fully realised within the specialist field of remote sensing. A delay of a few years is not uncommon between the implementation of new computer hardware and its acceptance by a specialist scientific user community. It may therefore be the early-1990's before the remote sensing community is widely utilising networked microcomputer workstations, with optical disk media for archiving and dissemination. Although this scenario is one of many which may be forwarded, it is one that could be implemented immediately with current hardware systems, and one which is becoming increasingly cost-effective.

It is apparent that advances in microcomputers and storage technology will increasingly satisfy the demanding system requirements of remote sensing scientists.

References.

Bernstein, R., Lotspeich, J.B., Myers, H.J., Kolsky, H.G., and Lees, R.D., 1984, Analysis and processing of LANDSAT-4 sensor data using advanced image processing techniques and technologies. *IEEE Trans. Geosci. Remote Sens.*, GE-22, 192–221 (1984).

Ferns, D.C., 1984, Microcomputer Systems for Satellite Image Processing. *Earth-Orient. Applic. Space Technol.*, Vol 4 (4), 247–254.

LANDSAT Data Users Notes, 1983, No.29, December. NOAA, Siuox Fall, S. Dakota, USA.

Townshend, J.R.G., 1981, The Spatial Resolving Power of Earth Resources Satellites, *Prog. in Phys. Geog.* Vol. 5 (1), 32–55.

7
A very low-cost Microcomputer-based Image Processor

P. J. Beaven
Transport and Road Research Laboratory
Department of Environment
Crowthorne
Berks. RG11 6AU

7.1. Overview.

The development of low-cost graphics systems means that a desktop image processor can be assembled from stock items. A system has been developed to display a 512 × 512 pixel image with sixteen levels of brightness in each primary colour. The software is written in BASIC to allow easy extension of the system, but calls machine code routines which write an image in less than 50 seconds. The interpreter is presented with a menu to run the system with a minimum of keyboard use. The main limitation on the widespread use of such instruments is the access to large data files; this is being solved by a direct link to a mainframe computer.

7.2. Introduction.

LANDSAT data provide a very useful source of information for engineering studies in developing countries. Regional studies may be made using simple photographic colour composites, but digital image processing is needed to extract the detail required from the smaller areas related to an individual project. Although much processing can be carried out on a standard computer, it is recognised that an interactive image processor is the most effective way of interpreting a LANDSAT scene. As the capital investment in such equipment is a minimum of £50,000, and more typically £100,000 or more, only a small number of research organisations and commercial companies have invested in them.

The situation has improved with the establishment of the National Remote Sensing Regional Centres in the U.K. and the purchase of systems by some Universities. However, this still means that in most cases the interpreter will not have access to the image processor in his own office, and many overseas countries that could benefit from the use of LANDSAT data will not have any facilities at all. Thus there still exists a need for a low priced image processing system that can be used in a local office.

The decreasing price of microcomputers, coupled with an increase in both their availability and sophistication, means that this need can now be satisfied. The widespread use of microcomputers has also led to the development of suitable peripherals which can enhance such a system. NASA have drawn up a specification for a Remote Image Processing System (RIPS) and simpler systems such as that described by Green (1981) have been available for some time. The Apple computer has also been widely used as described by Kiefer and Gunther (1983) demonstrating a need for a low-cost system to teach the principles of image processing. More recently the IBM personal computer has been used to demonstrate image processing and the FORTRAN programs published by Cracknell (1982) have been rewritten in BASIC for the Apple and BBC computers (see Myers and Bernstein, 1985). The National Remote Sensing Centre has also investigated the use of the BBC computer to demonstrate image processing in schools.

The use of the standard microcomputer screen limits the resolution of the image produced but the use of peripheral devices can circumvent this. Earliest examples were the use of a printer to produce images with greater information than could be displayed, but the availability of graphics controllers means that it is now possible to obtain a good quality, low-cost image. This chapter starts by describing the construction of an image processing system at the Transport and Road Research Laboratory using standard components. The object of this work was to establish the potential use of such an approach and to identify the main problems and limitations. Reference is also made to other solutions to the problem of providing low cost image processors. Potential users of such equipment include interpreters working in their own office in the U.K.; interpreters working overseas and educational institutions both in the U.K. and overseas.

7.3. Development of an image processor.

At the start of the programme, a broad list of objectives was drawn up, based on experience in using the facilities at the RAE Remote Sensing Centre and in developing a LANDSAT processing system based on an Intellect 200 at the Transport and Road Research Laboratory. These objectives included the resolution of the display, the priority list of functions to include in the program and the choice of operating system which in turn would influence the choice of computer.

When defining the quality of the display and the functions to be provided, it is necessary to consider the way the data will be interpreted. All types of digital processing are aimed at presenting the operator with an image that is easier to interpret. In some cases this is achieved by suppressing part of the data, thereby preferentially enhancing a desired feature. Operations such as density slicing or classification reduce in some way the amount of information that is displayed on the screen. In many geological investigations, and engineering surveys are most closely related to these, the objective is to present an image containing as much detail as possible so that the interpreter can apply his experience in predicting ground conditions. In this case an image processor should be able to alter the balance between different sets of data, but the display should be at the highest effective resolution.

Some preliminary trials were made using the IDP 3000 at Farnborough to preview an image made with a lower resolution display. The first test consisted of comparing a normal 512 × 512 image with the same area written using 256 lines containing 256 pixels. Although the lower standard produced a reasonable image of the area, there was a significant degradation in the quality of the image and so it was decided that a display resolution of 512 × 512 should be aimed for, even though this represented four times the amount of memory required to store the data.

In an attempt to reduce the memory required for the display, an investigation was made of the effect of cutting the colour resolution. It was found that an image with each band containing 16 levels of brightness was virtually indistinguishable from the original. When it is realised that the eye can hardly distinguish more than 32 levels (5 bits) in a single image, it is not surprising that a colour image occupying 3 × 4 bits, giving 4096 different colours, should contain as much information as the eye can perceive. Cutting the colour resolution to a total of eight bits, giving 256 colours, still gives a recognizable image, but with a very noticeable drop in quality.

Subsequent development of the system has shown that these broad definitions of image quality were realistic. The full image on the system, of 512 × 512 pixels with a range of 4096 colours, has proved very good for interpretation. It has also been found that an image containing 256 colours can be surprisingly useful, although showing obvious loss of detail. A comparison of these different display resolutions is shown in Figure 7.1 (see colour plates), which is a photograph of the colour monitor. The top half of the image is at the full 12 bit resolution. The lower half has been zoomed three times to display the difference between 12 bit and 8 bit colour resolution. Two colours can be displayed with eight levels but the third can only have four. The choice of colour to be restricted to four levels depends on the information in the scene; in this case it was found that restricting the blue gave the most acceptable image, but on other scenes restricting the red is most effective.

In summary it appeared that a system presenting a scene with 512 lines of 512 pixels, each pixel being displayed in 16 levels of red, green and blue, should give a high quality image for interpretation. The software would provide suitable routines to compress the data as it is written to the screen at full spatial resolution. As the compression of data would involve some loss of numerical accuracy, this meant that all further calculations would have to be made on the original data, and not use the information stored in the video display memory. Routines of this type would access data on the disc, and create a new file which would then be added to the display.

From these broad objectives the next stage was to identify equipment that could be used to build such a system. In keeping with the objective that this equipment should be in a local office rather than a specialist centre, it was decided that as many components as possible should be non-specialist, so that they could be used for other purposes. At that time, the only clearly established type of micro computer was CP/M based, and so it was decided that this should form the core of the system. It was envisaged that the system should be developed such that parts could be interchanged, and the software has been developed with this concept in mind. However the features of the chosen components have inevitably influenced the development of the system.

7.4. Equipment.

The central part of the system is the video RAM store which holds the data for the displayed image, and contains hardware to control the colours on the colour monitor. There are many graphic systems with a large range of colours which interface with minicomputers but this means that parallel ports are 16 bit wide and the prices tend to be high. The early microcomputer graphics systems generated 8 colours, but the situation has greatly improved with at least two UK companies offering micrographic systems with a resolution of 512 × 512 pixels and 4096 colours or more.

The Pluto system chosen for evaluation consists of a range of graphics controllers ranging from 8 to 24 bits/pixel. The memory is connected through 3 hardware look up tables to the colour monitor. Each look up table can be set to 16 values although these can be upgraded to give a choice of 256 levels. The display resolution is 576 interlaced lines each containing 640 pixels, although the normal image only occupies the central 512 × 512 pixel area. The internal command set includes a large number of line drawing and symbol commands. An important feature is the use of exclusive-or plotting which facilitates the implementation of cursor movement by drawing erasable lines within the video memory. More expensive equipment would use a separate bit plane to contain this small amount of line data. Connection to the host computer is either through an Z80 bus or parallel port, or interfaces can be provided for specific computers or RS232. The early models did not provide hardware zoom, but this feature can be replaced by a software zoom which rewrites part of the image. To complete the display a low cost high resolution colour monitor is needed and, as the display is interlaced, a long persistence phosphor is recommended.

The choice of a computer is closely bound up with choice of operating system and software. Although several 16 bit machines were available at the time, there was no dominant operating system. In addition most systems then available were only 8 bit externally and seemed to be little faster than the established machines. As the data processing involved 8 bit numbers, it was thought there would be little advantage in using the 16 bit machines, and so a search was made for a versatile CPM 80 based machine. The only other required feature was that it should be provided with a parallel port, as a serial port would be too slow for data transfer. The choice was influenced by other machines being used in the office, and a Torch with 20 Mb hard disc was chosen. The large storage capacity and extra speed of the hard disc is an obvious advantage, particularly when developing a system, but is not essential. One feature of the Torch is that although the programs are held in a Z80 controlled part of the machine, the input/output uses an internal BBC microcomputer, and as such the system could be adapted to run on a normal BBC computer, which could be an advantage for educational purposes.

7.5. Language.

Although a large number of image processing routines have been written in FORTRAN, this was not thought to be important for this equipment, as it was felt that considerable effort would be needed to implement them on a microcomputer. Jobs requiring that level of computing would best be carried out on a mainframe computer for display and enhancement on the micro system. It was felt that a combination of BASIC for general purpose routines, plus assembled machine code for data processing routines, would provide flexibility and speed where necessary.

As the required speed of processing was to be provided by machine code subroutines there was no need to consider using although being times faster in operation, is much slower when developing a program. On such a large program any minor change means a delay of several minutes while the program is recompiled, and when a program is interrupted it is not possible to examine the value of the variables to check the flow of the program. Microsoft BASIC is the most widely used interpreted BASIC and it can be compiled after the program is complete to improve its running speed.

Because of the format used to compact the data the standard Microsoft was slow in retrieving data, and so when BBC BASIC became available for the Z80, this was used instead. The built-in assembler and byte manipulation proved a great advantage for this system.

7.6. Program development.

The first decision made about the programming was that it should consist of a series of menus leading through the program, setting out a series of predetermined choices. This was in keeping with the concept that the system should be in a local office to be used by an interpreter who was not a regular user. The first stage of development was to devise a simple procedure to put an image onto the monitor and to modify the colour balance. The next stage was to add enhancements and the first of these was a software zoom. Figure 7.2 shows the initial menu which sets out the present level of development and shows the way the system can be used. The next stage of development has been to add various forms of spatial filtering as these techniques can be useful in emphasising linear geological features. No attempt has been made to consider classification techniques; density slicing could be easily implemented using the output look up tables, but as the data have been compressed this technique has less value than on a full resolution system.

```
T.R.R.L. IMAGE DISPLAY
Options available
1. See list of data files
2. Select source files
3. Display image
4. Set look-up table
5. Write single store
6. Edge enhancement
7. Subtractive ratio
8. Select display resolution
9. Zoom
```

Figure 7.2. Initial Menu Options.

One of the earliest decisions to be made in the programming was the format of data storage. A single band of data 512×512 contains 256k bytes which, if stored as ASCII characters, would need three times that space. For random access files on microcomputers it is normal to store data in strings using various software routines. On most 8 bit machines the maximum length of a string is 255 characters and so initially each line of data was divided into 4 strings containing 128 pixel values. The operating system uses the value of 13 to indicate the end of data and so pixels of this value cause problems. This was avoided by adding 128 to all values, as all original LANDSAT data are less than 127.

This structure was imposed by the way the BASIC language was used. Further development has shown that considerable increases in speed can be obtained by using the standard CP/M disc reading functions and that for this purpose it is better to organise the data into two strings each containing 255 pixel values. The reduction of line length to 510 is hardly noticeable compared with the significant decrease in time to read data. In this way it is still possible to use BASIC string handling for program development, but use the faster routine when writing to the screen. However if the CP/M function is included in a simple BASIC sub-routine, it is possible to have the advantages of BASIC for development with a full line length of 512 pixels; this also eliminates the need to add 128 to each pixel value.

The next stage was to determine the highest and lowest pixel values which would be used to calculate the compression function. In the early versions of development this was based on a 1% sample taken from the data file which gave an adequate sample. However to improve the speed of operation the complete histogram is now stored on the same disc as the data. Two methods are provided to determine the values determining the range of the data. The automatic method considers the distribution of data and sets the zero level of brightness to exclude 1% of the sample and sets maximum brightness to include 99% of the data. For many scenes this gives an acceptable image, but the colour balance can be adjusted by rewriting any or all of the bands. To do this the manual method of colour stretching is used. A histogram is displayed and a marker is moved with the cursor keys to identify the required values.

The final stage in creating an image is to read the data from the disc, obtain the compressed value from the software look up table and pass this value to the graphics controller which displays it on the screen. The early versions using BASIC disc handling took over 5 minutes to display a three colour image even when using a hard disc. By using more advanced machine code programming this has been reduced to less than 50 seconds and even using a floppy disc requires less than 5 minutes.

The same combination of BASIC and machine code is used to provide other routines including passing commands to the graphic controller, eg to draw the lines defining a cursor. The first enhancement provided was a software zoom, giving magnification by pixel replication. The next development was a BASIC subtractive ratio, which could also be used to provide a simple form of edge enhancement. The routine asks for the name of the the two files to be subtracted, and if they are different it assumes a ratio is required. If the two file names are identical it assumes that the enhancement routine is required and asks if the files are to be displaced vertically or horizontally. In either case a new file is created on the disc by subtraction of the input files. Using simple file handling and BASIC gave a slow program which took half an hour to process a file even when using the hard disc.

The images created in this way were very useful, but the time to produce them was unacceptable, so two file handling routines were developed to transfer the data from disc to memory. To provide ratios, two files are needed and so data is read in 8k blocks; for enhancement of a single file the data can be moved in 16k blocks. From these two routines it is possible to provide a series of ratios between files, or spatial filtering using a 3×3 window. To write a 512×512 file takes about 2 minutes from the hard disc and approximately twice as long when operating from a floppy disc.

7.7. Data transfer.

One of the main problems in the use of a micro-based image analyser is the access to the LANDSAT data which are normally provided on a full-size computer tape. Very few micros have facilities to read such tapes and so it is necessary to access a larger computer which can sample the data. Two years ago NASA proposed issuing subscenes on 8 inch floppy discs and on cartridges, and for the former they have standardised on the IBM format which can hold 4 banks of data, 256 pixels $\times$ 240 lines. As we had decided that this equipment should have a resolution of 512×512 the standard NASA format was not suitable. Using this format we could have stored our data on 4 separate discs to maintain compatibility. However, the majority of micros now use 5 inch micro discs, and in many cases do not support the large size. This tendency is being emphasised by the newer machines which are using even smaller discs.

The most common method to obtain data for a micro-computer is a direct serial line to a mainframe computer using a terminal emulation program. The general problem to be considered is the size of files to be transferred and the speed at which the equipment can operate to minimise the time involved. The simplest way to transfer the data is in hexadecimal format, i.e. one byte of data is represented by two ASCII characters between 0 and F. This results in a file of 512k for each LANDSAT band.

Speed of lines vary from site to site, but over long distances lower speeds are normal, and the rate used over telephone lines is 1200 baud. To transfer one band of data at this speed would take about one and a quarter hours. This time would be halved if the data were sent in transparent mode with each byte representing one pixel. The techniques to accomplish this will vary according to the local installation, which accounts for the widespread use of ASCII format. However, by use of such techniques and a faster line speed a file transfer in five minutes is possible.

7.8. Subsequent hardware development.

The micro computer industry is well known for the rapid rate of development, and since this project started there has been a change in the range of Pluto graphics controllers. The current range is based on multiples of an 8 bit board which effectively means that a system can have either 256 or 16 million colours. Hardware zoom is now provided and a frame grabber is available which can be used to digitise an image.

An 8 bit system could be used for an image processor concentrating on the display of classifications and enhancements. The quality of an image is adequate for the selection of training areas but is not really adequate for visual interpretation for which 12 bits seems to be the minimum. The full 24 bit system provides the basis for a very good image display system, and is used by several users, but there has been a considerable increase in cost when compared with the 12 bit system.

An alternative graphics controller is made by Digihurst, based on a six bit board. Such a system has been used by the Nature Conservancy Council to display LANDSAT data, and the boards are also incorporated in specialist equipment to provide an image display. Multiple systems can be supplied to give 12 bit systems, although there are as yet no references to such a system being used as the basis of an image processor.

7.9. Specialist systems.

An alternative solution to the provision of low-cost image processing has been developed at Preston Polytechnic under a contract from the National Remote Sensing Centre. This is not based on standard components, but consists of a specially designed system, called the LS10, which has built in software for image processing. As such it is not intended for alternative forms of computer processing.

The hardware comprises four banks of video memory, each capable of holding a $256 \times 256 \times 8$ bit image. Any memory bank may be connected through a look-up table to the colour monitor. There are also three separate $256 \times 256 \times 1$ bit overlay planes which can be used for histogram plotting or image classification.

Operation of the system is controlled with a purpose made panel containing a joystick and extended numeric keypad. The monitor either displays the image or the menus used to select the required function. The main facilities available are:

1) Zoom. Hardware zoom of *2 or *4. The joystick is then used to pan over the image.

2) Contrast stretch. Automatic or manual contrast stretches are used to program the look-up tables and adjust the colour balance.

3) Copy. Copy one memory bank to another, with or without the transformation applied by the look-up table.

4) Histogram. Displays a histogram in an overlay plane.

5) Density slice. Allocates a particular colour to a selected range of image intensity values.

6) Classify. Box classifier based on training areas identified on the image.

7) Filter. Low- or high-pass filtering of an image. 8) Arithmetic. (+, −, × and ÷) between two images.

The standard system obtains data from a host computer via a serial link, but data can be stored on the optional floppy disc unit. The standard disc format will hold 12 images and data can be obtained in this format from the National Remote Sensing Centre.

The LS10 provides a low cost image processor with a good range of basic software. The main difference to the user, when compared to the system described above, is the spatial resolution of the displayed image. On a colour monitor a 256×256 image is obviously composed of rectangular pixels whereas a 512×512 image appears to be a normal picture. In addition to the improved visual appearance, the 512×512 image displays four times the area. This ability to see a larger area in one image is a significant benefit when making an interpretation.

The advantages of displaying a smaller image are not so readily apparent to the user, but need to be remembered. The main advantage is the speed with which an image is processed and displayed, using a floppy disc, not the very much faster hard discs. The second advantage is that by restricting the size of the memory, it has been possible to produce a complete system at a very competitive price.

7.10. Conclusions.

Standard micro-computer components can be used to provide a low cost image processor giving a high-quality image that can be used for interpretation. It is practicable to obtain a spatial resolution of 512×512 pixels on the screen; lower levels of resolution are significantly inferior. Economy of video RAM storage can be made by compressing the data to show a maximum of 4096 colours as this is approaching the limit of the eye's discrimination. Many image processing techniques consist of simple manipulations applied to large amounts of data. It has been found that the major part of the programming can be easily implemented in BASIC which calls a small number of short machine code routines for data processing. In this way a full colour image can be produced in less than one minute and simple forms of edge enhancement can be performed. This equipment has been designed to be used as a work station linked to a large computer which extracts the data from the LANDSAT tape, and could process it before display. Speed of transfer to the work station will vary according to the line rates and data format. A typical time for four bands of data would be 20 minutes, but using public telephone lines this would rise to over 6 hours. This demonstrates the need for another method of data transfer using transportable media.

Acknowledgements.

Crown copyright: 1985. The work described in this Chapter forms part of the programme carried out for the Overseas Development Administration, but any views expressed are not necessarily those of the Administration.

References.

Cracknell, A.P., 1982, *Computer programs for image processing of remote sensing data*, University of Dundee 1982.

Green, K.M., 1981, Digital processing of tropical forest habitat in Bangladesh and the development of a low cost processing facility at the National Zoo, Smithsonian Institution. Proc. 15th Int. Symposium on Remote Sensing of Environment pp 1315,1326. *ERIM*, Ann Arbor.

Kiefer, R.W. and Gunther, F.J., 1983, Digital processing using the Apple-II microcomputer. *Photogrammetric Engineering and Remote Sensing.*, 49 (8) pp1167–1174

8
Capturing Image Syntax using Tesseral addressing and arithmetic

S. B. M. Bell, B. M. Diaz and F. C. Holroyd
NERC Computing Service
Holbrook House
Station Road
SWINDON SN1 1DE
Wiltshire, ENGLAND

8.1. Introduction.

We consider an image to have syntactic components and semantic components. However, given that an image is part of the plane, and given that we can conceive the physical image as a collection of "pixels"—what constitutes the image syntax? More importantly can we capture the elements of this syntax implicitly in a formalism which can be implemented on a computer? This Chapter concentrates on these aspects of image understanding and proposes a solution which the authors believe goes some way to answering such questions.

Arguably the most important syntactic entities of the physical image are pixel shape and pixel neighbourliness. A crucial concern with these two is that they should be scale independent. In other words the shape of a pixel and its neighbourlinesss should follow well formed rules which govern their relationship one to another at any scale level, and between scale levels. Furthermore, this hierarchical relationship should at the same time cover the entire Euclidean plane and yet admit of localised area coherence.

At this point, we note that computer considerations such as data storage efficiency and problems of data retrieval should play no part in our deliberations. However, in assessing the validity of any approach, its amenability to computer implementation and relative efficiency are key factors. Furthermore, it is often the case that approaches designed for computational efficiency have explanatory value because of their simplicity.

8.2. Tesseral Addressing and Arithmetic.

Tesseral addressing and arithmetic was first proposed as a solution to the problems of image syntax by Bell *et al.*, (1983,1985) and was based on work done by Holroyd (1983). That work stemmed from an interest in using hierarchical data structures as a means of storing cartographic data and in particular an interest in the linear quadtree proposed by Gargantini (1982), although we note that similar work was done independently by Woodwark (1982) and Oliver and Wiseman (1983). Antecedents for tesseral addressing and arithmetic are to be found in the work of Lucas (1979) and Gibson and Lucas (1982). An excellent review of quadtrees and their many ramifications is presented in Samet (1984).

We will show that replacing cartesian methods of image processing with tesseral methods better captures the syntax of an image and also provides for more efficient computation. More particularly, a rule-based arithmetic complements the tabular approach Bell *et al.*, (1983) and in one particular case, the 4-shape over the 4^4 tiling, can be implemented easily in hardware. Finally, we introduce the concept of an ordering, the tesseral raster, which preserves area coherence and which as the bottom-up-quadtree has been described by Rosenfeld and Samet (1979) and which has been the subject of storage efficiency studies (Morton, 1966; Goodchild and Grandfield, 1983) and data insertion analyses (White, 1982).

8.3. Quadtree Storage.

The quadtree is an hierarchical data structure in which an image can be held. The image can be built up from the pixels of the 4^4 pixellation of the plane or indeed the 3^6 pixellation (See Bell *et al.*, 1983) whence it is known as the bottom-up-quadtree. Alternatively, we can think of an image successively quartered into the top-down-quadtree (figure 8.1).

By recognising areas of homogeneity in the image (image semantics?) the quadtree seeks to maximise area coherence rather than the raw linear coherence afforded images held in raster form. However, we shall not be seduced by the elegance of the quadtree approach and merely note that the top-down-quadtree **begins** with an image, which implies that it cannot easily be applied to the entire Euclidean plane in the way that we require. Furthermore, the anathema of mixing image semantics and worrying about computational efficiency so early in the discussion rule out such an approach being used as the basis of syntax. However, the bottom-up-quadtree requires merely the unique location of some point and an operational direction in which to work. Providing these are available, the entire Euclidean plane can be addressed and the explanatory simplicity of the quadtree idea exploited.

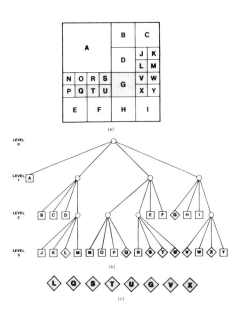

Figure 8.1. An example image and its quadtree representation
(a) An image is decomposed for quadtree representation by successive quartering. Thus, after the first quatering "A" needs no further decomposition, while the other 3 quarters do require further decomposition. When no further quartering is required the image is said to be top-down decomposed.
(b) An "explicit representation quadtree" labels each data leaf (designated by squares and lozenges) and stores data with every label.
(c) Implicit representation, labels the "black" leaves with the size of the region represented and maintains the order of these leaves. Because the original size of the image is known, the complete image can be reconstructed by interpolation from the implicit representation.

8.4. Tesseral addressing of Quadtrees.

Gargantini (1982) showed how a quadtree could be stored efficiently (avoiding tree pointers) in a linear fashion as a sorted list of integer addresses. These addresses were an implicit encoding of the path from the quadtree root to the data leaf, which had the property that those addresses that were numerically close together were also spatially close together. Bell *et al.* (1985) generalised the addressing structure such that the tesseral addresses were uniquely located in the Euclidean plane. It was also shown that such addressing could be applied to other hierarchical areal organisations.

The association between the data bearing leaves of the quadtree in Figure 8.1b and tesseral addresses can be achieved in many ways. The simplest of these is the Gargantini list, in which only the 'black' leaves are considered (Figure 8.1c). If we use the digit 4 in the role of a wild-card which can be replaced by any of the valid tesseral digits (0,1,2,3) then the address of leaf A in Figure 8.1a is 244 if it occurs in the positive quadrant (cf Figure 8.2a).

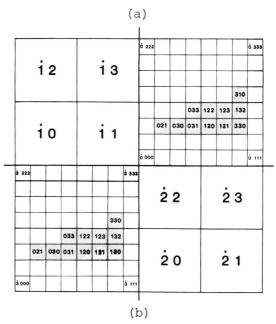

Figure 8.2. Tesseral addresses used to label quadtree leaves
(a) A tesseral grid contains four quadrants labelled $\vec{0}$ (the positive quadrant), $\vec{1}$, $\vec{2}$ and $\vec{3}$ (the negative quadrants). The figure shows the same "object" placed in the same relative position in the $\vec{0}$ and $\vec{3}$ quadrants. (Only key tiles of the tesseral grid have been labelled—it is left as an exercise for the reader to label the remainder).
(b) Implicit representation of the "object" is located on the tesseral grid by means of the use of the dotted digit (not illustrated). The size of the object is determined by the number of digits in the non-dotted portion. To halve the size of the displayed object a leading 0 is added and that set of addresses used on a finer grid. The 12x symbol indicates the four values of 120, 121, 122 and 123. The x could be implemented as the digit "4".

When this list is sorted it may be seen (figure 8.2b) that addresses have a tendency to be close together in the list if the corresponding leaves are close together in the image. The list of addresses can obviously be used as labels by which colour or other data such as texture could be stored.

If the whole image had been double in size then four tesseral digits would have been necessary for each pixel address stored. However, the pixel addresses in figure 8.2b would only change by the addition of a leading 0. Leading 0s can be ignored as is ordinarily the case, however, in tesseral addressing they should never be forgotten. The reason for this stems from the value the concept of an infinite number of leading 0s provides in tesseral addressing. We symbolise this infinite number of leading 0s with $\dot{0}$ (the dot being identical in value to the dot used to indicate recurring digits in decimal fractions) and to avoid confusion we place a printer's en space between the 0 and the rest of the address. Because of the possibility of digits recurring to the left and right we identify tesseral recurring digits with an arrowhead e.g. $\vec{2}$. Left recurring or tesseral digits are described as being left beaked or beaked digits and thus $\vec{2}$ 3 is beak (or left beak) two three. Figure 8.2a illustrates how the image may be overlaid in the other quadrants providing that beaked tesseral digits (cf negative cartesian co-ordinates) are used instead of ordinary positive addresses (ones preceded by $\dot{0}$). Clearly if only the positive quadrant is being discussed, then the dot can be ignored altogether, although care must be taken if transformations which straddle the quadrant boundaries are considered. The recurring digit symbolism is reminiscent of the twos-complement method of holding a negative number in computers and is the basis of the discussion in Section 8.6.

8.5. Tesseral Raster Storage.

The benefit of holding an image in quadtree form increases as the homogeneity of the image increases. However, if remotely sensed images are considered as an example, it must be recognised that they have little spectral homogeneity and in fact require extra storage over traditional raster storage if held faithfully in quadtree form. We note that for classified or otherwise processed images this is less so and quadtree storage may be more appropriate in these cases.

Tesseral arithmetic may still be applied to faithfully held remotely sensed images by use of tesseral raster storage format. The normal image may be thought of as a linked list in which x varies uniformly between adjacent data items, and periodically at the end of raster lines it varies in y (figure 8.3a). If tesseral addresses were used rather than cartesian addresses (those using x and y "coded" in some way) then the pixel sequence of normal raster images would be 000 001 010 011 100 111 002 003 012 013 102 103 112 113 20 21 31 ...(figure 8.3a), assuming the image was decomposed from the bottom upwards as in figure 8.2a. The tesseral raster attempts to improve on this by sorting the list into the order 000 001 002 003 010 011 012 013 020 021 ...(figure 8.3c).

Although we have taken the 4-shape generated from the 4^4 tiling as our example, it is clearly the case that the tesseral raster is a general device and

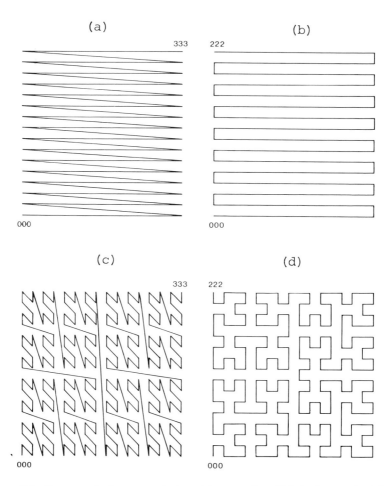

Figure 8.3. Four standard raster orderings (a) row order. This is the standard "raster" decomposition, (b) row-prime order, (c) N (or Morton) order. This has also been called the top-down-quadtree ordering or decomposition, (d) pi-order.

is dependent only on there being an order in which the atomic tiles in any molecular tile are taken. As with the normal raster form these addresses may be implicit, since the position of the pixel in the list is its position to the base 4. Storage on disc and on sequential media such as magnetic tape could take advantage of this format. Alternatives such as row-prime and pi-order rasters (figures 8.3b and 8.3d) could also be used, however, the advantages of a sorted list would be lost, with little gain in efficiency (Goodchild and Grandfield, 1983).

It may be seen that the storage required for the tesseral raster is identical with that required for the normal raster. However, if techniques such as run-length encoding were used, area homogeneity could be exploited and considerable storage efficiencies could be made (Goodchild and Grandfield, 1983). It may also be seen that the run-encoded tesseral raster is very nearly the Gargantini list of quadtree leaves, mentioned above, with the difference that the size of each leaf is stored rather than the quadtree address. Conceptually, it may be converted into a quadtree by insisting that each run be divided into the smallest number of runs which could be given a quadtree leaf address.

8.6. Tesseral Arithmetic.

The importance of tesseral arithmetic lies in the possibility of mapping spatial operations onto arithmetic operations. For example, in the case to be described, addition produces a translation and multiplication results in a scaling and rotation. In alternative tesseral systems, derived from the same 4-shape 4^4 tiling and addressing, such geometric interpretation for the arithmetic operations are sometimes not possible. It is interesting to note that where different geometric interpretations are possible, these may produce novel or unexpected transforms. Although such systems may have special application in image processing, they are as yet not fully investigated. The quadtree-based arithmetic under consideration is in design similar to ordinary arithmetic in that it has a place system with carry. Once the result of the addition of all possible pairs of tesseral digits is known (Figure 8.4a), the addition of any two tesseral numbers, for example tesseral 312 and 120, can be performed (figure 8.4c). After adding a column of digits, the rightmost digit of the answer is entered as the result for that column and any surplus digits on the left are carried on the the next column or columns. This method also applies without change to the negative, beaked digits.

The familiarity of this behaviour does not extend to the answer produced when any pair of digits is added! For example tesseral 2 plus tesseral 3 is tesseral 21 not 5 as might have been expected. The addition table is generated by application of the rule that addition of any number to zero produces a conversion or translation of zero to that number. In symbolic terms:

$$0 + 3 \to 3 \qquad (8.1)$$

In general terms addition of any number to 0 is a move from the address (tile) 0 to the address (tile) represented by that number (figure 8.4b). Thus addition anywhere on the plane can be represented geometrically by a move identical in direction and scale to the addition of that number to 0. It may

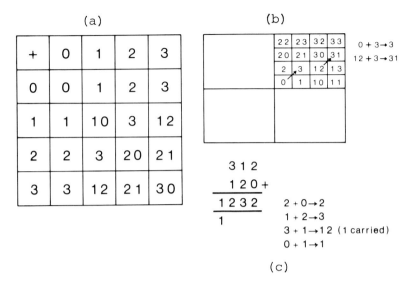

Figure 8.4. Tabular tesseral addition. (a) The tesseral addition table, (b) Generation of the addition table is based on the observation that translation of the tile 0 to the tile 3 is the addition 0 + 3. Thus, the addition anywhere on the tiling (e.g. 12 + 3) is a similar translation (i.e. to 31 in the example). (c) A worked example of the addition of two tesseral addresses which use the addition table.

be seen that the reverse process, addition of 0 to any number is a zero translation from the addressed point (tile) represented by that number—that is:

$$3 + 0 \rightarrow 3 \qquad (8.2)$$

For a more complete description of tesseral addition see Bell *et al.*, 1983. Tesseral subtraction can be thought of as a reversal of the addition operation, in other words:

$$3 - 3 \rightarrow 0 \qquad (8.3)$$

that is, subtraction of 3 is a move backwards from address 3 to address 0. In general, subtraction involves generating the complement of a tesseral address, using a complement vector (Figure 8.5a) and then (tesserally) adding the complemented address.

A worked example of tesseral subtraction is presented in figure 8.5b, in three steps.

In tesseral multiplication 0 and 1 play their usual role, so that multiplication by 0 reduces any tesseral number to 0 and multiplication by 1 leaves it unchanged (the identity element). Furthermore we ensure that the multiplication is commutative such that, as with addition, the order in which a

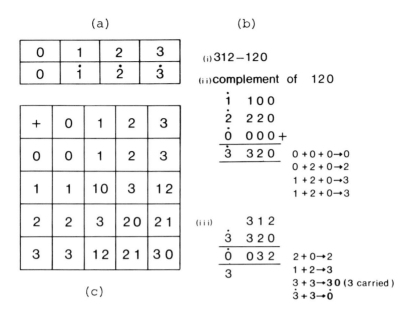

Figure 8.5. Tabular tesseral subtraction. Subtraction is performed in two steps, the generation of the complement vector in a and then an addition of the complement as illustrated in b(iii). (The addition table (c) is reproduced to aid the reader)

multiplication is performed is irrelevant. Thus we have that:

$$1 \times 3 \to 3 \quad \text{and} \quad (8.4)$$
$$3 \times 1 \to 3 \quad (8.5)$$

If the tesseral number 1 is thought of as the vector from 0 to 1, multiplication by 3 results in an anticlockwise rotation of the vector $\underline{01}$ through 45 degrees with a $\sqrt{2}$ scaling, to generate the vector $\underline{03}$ (figure 8.6b).

In general terms a multiplication by $\overline{3}$ anywhere is an anticlockwise rotation of 45 degrees and a scaling of $\sqrt{2}$. A worked example of multiplication is presented in figure 8.6c. (To aid the reader the normal addition and multiplication tables have been extended to include the negative, beaked digits—figures 8.6a and 8.6b respectively).

Tesseral division is performed analagously to ordinary decimal long division in that a multiple of the divisor is subtracted from a portion of the dividend such that the remainder has fewer digits in it than had the portion of the dividend. The difference lies in that the non-zero multiples of the divisor are equivalent in size, and it is only by trial and error that a

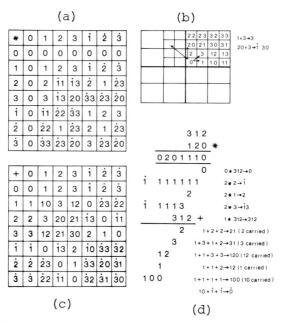

Figure 8.6. Tabular tesseral multiplication. (a) The tesseral multiplication table (extended by rotated digits) (b) generation of the multiplication table is based on the observation that rotation and scaling of vector $\underline{01}$ to $\underline{03}$ is a multiplication by 3. Thus multiplication anywhere on the tiling (e.g. 20×3) is a similar rotation and scaling (i.e. to $\vec{1}\,30$). (c) A worked example of multiplication. Note that all the individual multiplications are performed first then the final addition is performed. Those experienced can of course skip this intermediate step. (d) result of tesseral multiplication.

suitable multiple is found. Furthermore, there are 6 multiples because both the positive and beaked digits must be considered. Thus, in the worked example (figure 8.7c) the complement of the multiples of the divisor are first generated (figure 8.7b) and then division proper is performed.

It may be noted that the choice of which digit to place in the answer is sometimes complicated by having more than one possible answer, as happens in figure 8.7c where both "3" and "1" have fewer digits than "20", allowing the answer to contain either 2 or 3 in the first column. The effect of the 'wrong' decision is to complicate the removal of embedded beaked digits (figure 8.7c(iii)) although the final answer produced is the same in both cases. A similar situation occurs in ordinary divison, if the 'wrong' choice of digits is made. However, in ordinary division the choice of digit is made simpler because of the predefined order $0 < 1 < 2 < 3 < 9 \ldots$ However, for quad tesserals the order $0 < |1| = |2| < |3|$, which has been found empirically by Diaz (1985) and which may have an axiomatic basis

(a) $\dfrac{201110}{120} = \dfrac{20111}{12}$

(b)
$-(1 \times 12) = -(12) = 1\dot{\ddot{2}} = \dot{3}\,2$
$-(2 \times 12) = -(\dot{1}31) = 1\dot{\ddot{3}}\dot{1} = \dot{2}\,1$
$-(3 \times 12) = -(23) = \dot{2}\dot{\ddot{3}} = \dot{3}\,13$
$-(\dot{1} \times 12) = -(\dot{3}2) = 3\dot{\ddot{2}} = 12$
$-(\dot{2} \times 12) = -(\dot{2}1) = 2\dot{\ddot{1}} = \dot{1}\,31$
$-(\dot{3} \times 12) = -(\dot{3}13) = 3\dot{1}\dot{\ddot{3}} = 23$

(c)

(i)

1	2	3	$\dot{1}$	$\dot{2}$	$\dot{3}$
2 0	2 0	2 0	2 0	2 0	2 0
$\dot{3}$ 2	$\dot{2}$ 1	$\dot{3}$ 1 3	1 2	$\dot{1}$ 3 1	2 3
$\dot{2}$ 1 2	1	$\dot{3}$ *	3 2	$\dot{1}$ 3 1 1	2 0 3
$\dot{3}$ 1	$\dot{3}$ 1	$\dot{3}$ 1	$\dot{3}$ 1	$\dot{3}$ 1	$\dot{3}$ 1
$\dot{3}$ 2	$\dot{2}$ 1	$\dot{3}$13	1 2	$\dot{1}$ 3 1	2 3
$\dot{3}$ 0 3	2 0 0	$\dot{3}$132	$\dot{2}$ 3	$\dot{1}$ 0	2 *
2 1	2 1	2 1	2 1	2 1	2 1
$\dot{3}$ 2	$\dot{2}$ 1	$\dot{3}$ 1 3	1 2	$\dot{1}$ 3 1	2 3
$\dot{1}$ 3	1 0	$\dot{2}$ *	3 3	2 0 0	2 1 2
$\dot{2}$ 1	$\dot{2}$ 1	$\dot{2}$ 1	$\dot{2}$ 1	$\dot{2}$ 1	$\dot{2}$ 1
$\dot{3}$ 2	$\dot{2}$ 1	$\dot{3}$13	1 2	$\dot{1}$ 3 1	2 3
$\dot{3}$ 1 3	$\dot{2}$ 1 0	$\dot{2}$02 2	$\dot{2}$ 3 3	0 0 0 *	1 2

(ii)

```
            3 3̇ 3 2̈
     12 ⟨2̇0⟩1 1 1
          3̇ 1 3    +
         ⟨3̇1⟩1 1
            2 3    +
           ⟨2̇ 1⟩1
            3 1 3  +
             ⟨2̇ 1⟩
             1̇ 3 1 +
             0 0 0
```

(iii)

```
  .  3 0 0 0
  3̇ 3 3 3 0 0
       3 0
  2̇ 2 2 2 2 2 +
    0 0 3 1 2   0+2→2
          2     3+2→21 (2 carried)
          2     3+2+2→23 (2 carried)
        2 1     3+3+2+2→210 (21 carried)
        3       3+2+1→30 (3 carried)
                3̇+2̇+2+3→0
```

Figure 8.7. Tabular tesseral division. (a) we note that 20110 divided by 120 is equivalent to 2011 divided by 12. (b) Division proceeds in two steps. The generation of the multiples of the divisor (e.g. 12). These are complemented and used in the frame in c(i). (c) A suitable portion of the divident is taken (the digits in the ovals). To these the multiples generated in (b) are added. Two possible multiples (of 2 and 3) provide a remainder with fewer digits than the portion. One of these (starred) is chosen arbitrarily. The remaining digits (iii) are then "brought down" and the process repeated. Finally, the answer with embedded dotted digits is worked out c(iii).

along the lines suggested by Bell (1985a) ensures that division proceeds in a tidy manner. In the example we have avoided a "fractional" answer.

Tesseral fractions do exist (and indeed constitutes the power of the system as a whole) but are not discussed here. The interested reader is directed instead to Diaz, 1985 for further details.

8.7. Rule-based tesseral arithmetic.

The rule-based approach to tesseral arithmetic was developed as an alternative to tabular methods, for aesthetic as well as computer implementation reasons. We note that the approach is inherent in the Gargantini, 1982 algorithm for inter-converting cartesian and linear quadtree co-ordinates, and as such has been the subject of investigation for hardware implementation. The method makes use of the representation of a tesseral address as two binary numbers. A possible hardware implementation can be anticipated by referring to these two numbers as the T_x and T_y registers of a tesseral computer. The conversion of a tesseral number into its register representation is done in two steps. Each digit of the tesseral address is represented in its binary form working left to right, thus tesseral 2321 becomes 10 11 10 01, then the first digit from each pair is placed in the T_y register and the second digit is placed in the T_x register. This results in T_x containing 0101 and Ty containing 1110. Figure 8.8 describes the process diagrammatically.

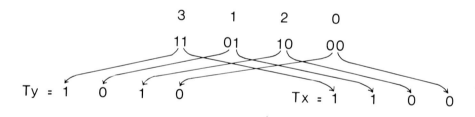

Figure 8.8. Tesseral to Cartesian address conversion. Tesseral to Cartesian address conversion is performed by considering each successive tesseral digit in its binary form and then partioning the result such that the first digit of each pair goes to form the y component and the second digit the x component.

Rule-based tesseral addition is accomplished by performing two's complement binary addition separately in each of the registers and recombining the separate halves by a reverse of the process in figure 8.8. Subtraction, is either the reverse of addition using the binary registers or may be done be generating the tesseral complement and performing addition as above.

Tesseral rule-based multiplication is performed with four rules, one for each of the tesseral digits:

Rule 1: Multiplication by 0.
Both T_x and T_y registers are set to 0

Rule 2: Multiplication by 1.
Both T_x and T_y registers are left unchanged.

Rule 3: Multiplication by 2.
Taking the binary digits in the T_y register one by one and working from right to left add (binary) beak 1 in the corresponding position in the T_y register for every 1 that occurs (this process is called tesseral register complementation). Then swop the contents of the T_x and T_y registers to produce the tesseral result illustrated in figure 8.9b. (In figure 8.9a, the same result is obtained by tabular means).

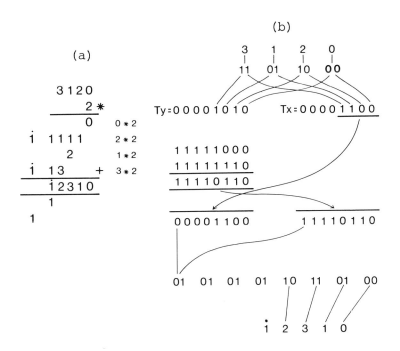

Figure 8.9. Rule based multiplication by 2. (a) For confirmation the tabular multiplication. (b) Rule based multiplication by 2 is performed in 4 steps. Binary decomposition as described in Figure 8; binary complementation of the T_y register by addition of 1 into the position occupied by 1s in the register; register swopping and recombination to give the tesseral result.

Rule 4: Multiplication by 3.

Multiplication by 3 can be derived by considering that 3 is (2 + 1). Consequently, 3120 multiplied by 3 is $3120 \times 2 + 3120 \times 1$. The addition is done by adding the original contents of T_x and T_y to the contents after execution of rule 3.

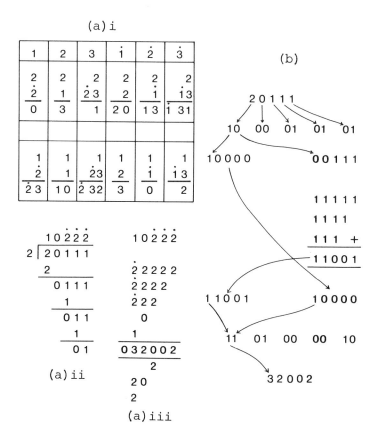

Figure 8.10. Rule based division by 2. (a) For confirmation the tabular division by 2. (b) Rule based division by 2 is identical to multiplication with the exception that the T_x rather than the T_y register is complemented.

Rule-based division by the individual tesseral digits has similar rules to those for multiplication and consequently only the rule for multiplication by 2 will be exemplified. The binary decomposition is as before; however, to divide by 2 we complement the T_x instead of the T_y register. The registers are swopped and the final tesseral value of the division is obtained by register recombination (figure 8.10).

At the time of the first draft of this Chapter (presented to the Remote Sensing Society in 1984) the authors had not devised "neat" algorithms for rule-based multiplication and division by multi-digit tesseral numbers and offered a bottle of champagne to the "neatest" coded solution (in Pascal) received by May 1st, 1985. Because of the decision to publish the paper in book form, that draft did not see much of the light of day and consequently although the authors now have "un-neat" solutions, the "neatest" solution received by May 1st 1989 will receive the maturing champagne. (The authors shall be the sole arbiters elegantiae regarding "neatness" and their decision shall be final).

This methodology has not been extended to performing these operations automatically because the authors are not experienced in hardware design techniques or in the intricacies of this type of logical operation. Suffice it to say that the tabular and rule-based approaches have been coded and give consistent results; a hardware implementation is eagerly anticipated and the authors may be contacted for enthusiastic, though amateur help and advice.

8.8. Use of Tesseral Methods in Image Manipulation.

In previous sections, the transformations induced by tesseral arithmetic were described. If tesseral addresses could be generated quickly from the storage format or if they were explicitly stored, then these could be manipulated to provide image transformations. For example "panning" across a large image (or image database?) is performed by invoking tesseral addition to translate the image. Tesseral multiplication is used for the combined operations of scaling and rotation. Combinations of these basic operations can be used to achieve all of the standard image manipulation operations. Furthermore, because the tesseral address is related to region size, image samplings can be performed to provide a low resolution image or a database "browse" capability.

Although we have described tesseral addressing using the concept that an address, point and tile are synonymous—there are problems with such a simplification. The point/tile duality problem occurs because of the confusion of semantic and syntactic properties. The region, or at this level, the tile consists of "tiles" at a lower level—because of our insistence on scale independence. Thus, to recognise one set of tiles while ignoring a lower set of tiles is to break the rules. Although for practical (semantic) reasons this may be necessary the tesseral system *per se* does not have the problem. For image processing purposes we can define the exact point addressed by n-tesseral digits as commensurate with whatever level of accuracy is required in our calculations Bell *et al.* (1985) and deal with the inconsistency only when greater accuracy is required. Thus, a tile (one address) can represent the straight line from the bottom left corner of the tile to the top right. At a higher level of accuracy 2 addresses are needed (figure 8.11) to describe the same line. (Clearly a line within the terms of this paper is a semantic unit). This ability to consider an address as either a point or tile is a strength not possessed by the cartesian system where every address (x and y) must be a point (or have error attached to them) and must have zero dimension (accuracy).

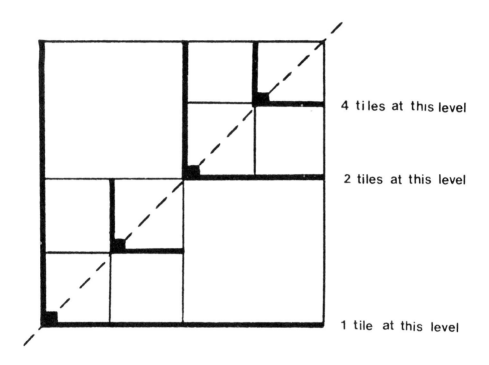

Figure 8.11. Representing accuracy. A tile (one address) can represent the straight line from its bottom left corner to its top right. The same line at a higher level of accuracy needs two tiles/addresses. This ability to consider an address of a point as a tile at some level of accuracy is a strength of the tesseral system.

Other image manipulations which we have investigated and which are potentially faster if a tesseral approach is taken are affine and fourier transformations. However, these are not discussed here; the reader is advised to turn to Bell, 1985b for further details.

8.9. Summary and Conclusions.

We have suggested the idea that tesseral addressing and arithmetic can be used in all aspects of image processing. We propose that the primary reason for this is that the tesseral system better captures image (and graphical?) syntax than the cartesian system—although it suffers the problem that it is a later invention. It is also the case that a tessseral structure is as good as, if not better than a cartesian for computer implementation. This is because tesseral addresses can be used to label the leaves of storage structures (e.g. quadtrees). More importantly tesseral addresses offer all the advantages of an hierarchical formulation, without the need for that hierarchy to be the one used for data storage. For example a B-tree (Bayer and McCreight, 1972) coupled with a tesseral raster ordering provides scope for preserving area coherence and simultaneously providing rapid access and data compression. Clearly much work is required to categorise all the advantages of the use of tesseral systems, however, its future seems bright.

We have shown how tesseral arithmetic provides further advantages and can be used to perform most if not all the standard image manipulation operations. Although considerable work needs to be done to devise efficient algorithms for these operations, it would appear that the approach is sound and in certain areas, provides significant speed improvement over conventional approaches. We also note that the rule-based approach is ideally suited for hardware implementation suggesting the possibility of extremely rapid, real-time processing of complex and voluminous image data.

Open research issues which remain and which it is our intention to address are those concerned with higher dimensions, and as a special case the surface of the sphere. Pragmatic issues which need further thought concern the rule-based approach and the development of basic operations which can be implemented in hardware. Clearly a methodology of assessing the performance of all these algorithms is also needed, especially one which allows comparisons with conventional methods. The development of image processing algorithms which utilise the tesseral approach is also seen as being extremely important and some work in this direction is already underway. Finally, there are many database issues which a tesseral approach brings into focus and it is perhaps these areas which require the most urgent attention.

Acknowledgements.

We thank the Natural Environment Research Council (NERC) for making available the facilities and resources for this research. We would also like to acknowledge the painstaking work done by Elizabeth Farnell in preparing the illustrations for publication.

References.

Bayer R. & McCreight E.M., 1972, Organisation and maintenance of large ordered indices. *Acta Informatica* 1(3) 173–189.

Bell S.B.M., 1985a, Constructive tesseral algebra. *Proceedings of the First Tesseral Workshop*. (To appear).

Bell S.B.M., 1985b, Some tesseral algorithms. *Proceedings of the First Tesseral Workshop*. (To appear).

Bell S.B.M., Diaz B.M., Holroyd F.C. & Jackson M.J., 1983, Spatially referenced methods of processing raster and vector data. *Image and Vision Computing*, 1(4); 211–220.

Bell S.B.M., Diaz B.M. & Holroyd F.C., 1985, Tesseral addressing and arithmetic of quadtrees. (To appear)

Diaz B.M., 1985, Tabular tesseral methods and calculators. Proceedings of the First Tesseral Workshop. (To appear).

Gargantini I, 1982, An effective way to represent quadtrees. *Communications of the Association for Computing Machinery*, 25(12); 905–910.

Gibson L., & Lucas D., 1982, Vectorisation of raster images using hierarchical methods. Computer Graphics and Image Processing, 20; 82–89.

Goodchild M.F. & Grandfield A.W., 1983, Optimising raster storage: An examination of four alternatives. In: *Procs of the 6th International Symposium on Automated Cartography*. Ed. B.S. Weller, pp 400–407.

Lucas D., 1979, A multiplication in N-space. *Procs of the American Mathematical Society*, 74(1); 1–8.

Morton G.M., 1966, A computer orientated geodetic data base and a new technique for file sequencing. *Internal report for IBM Canada Ltd*, Ottawa, Canada.

Oliver M.A. & Wiseman N.E., 1983, Operations on quadtree en- coded images. *The Computer Journal*, 26(1); 83–91.

Rosenfeld A. & Samet H. (1979) Tree structures for region representation. *Proc. 4th Int. Symp. on Computer Assisted Cartography*, (Auto-Carto IV); 108–118

Samet H., 1984, The quadtree and related hierarchical data structures. *ACM Computer Surveys*, 16(2); 187–260

White M., 1982, N-Trees: Large ordered indexes for multi-dimensional space. *Internal Report*, US Bureau of the Census, Washington DC 20333

Woodwark J.R., 1982, The explicit quadtree as a structure for computer graphics. *The Computer Journal*, 25(2); 235–238

9
Multiple source data processing in remote sensing

P. T. Nguyen and D. Ho
IBM Scientific Center
36 ave Raymond Poincare
75116 Paris, France

9.1. Introduction.

Digital image processing in remote sensing started with the launch of the series of LANDSAT satellites. LANDSAT data were then considered as the "standard" for many applications of remote sensing.

Recently other sources of digital data, either from other types of satellites, the shuttle or from the digital conversion of analogic data (topographic, geophysic, aerial photographs etc.), provide users with a considerable amount of useful information for earth science applications. These data are heterogeneous in their format, geometric and radiometric properties, and temporal sampling rate. Hence the integration of these data into an operational data bank will require both the data format conversion and the geometric and radiometric corrections. Their exploitation will require digital image processing, pattern recognition techniques, and modelling.

In this chapter, techniques of processing multi-source data are presented with their applications in geology and climatology. The data used consist of satellite data from LANDSAT, HCMM (Heat Capacity Mapping Mission), NOAA-AVHRR and METEOSAT (European Meteorological Satellite) and non-satellite data such as DTM (Digital Terrain Model), gravimetric measures, and digitized geological maps.

9.2. Data Correction.

Geometric Correction.

Satellite images are subjected to different deformations due to the Earth, the satellite, the orbit, and the image projection. The contribution of the Earth comes from its rotation, oblateness, and curvature. The satellite contributes to the image deformation by its variation in speed, attitude, and altitude. The scan skew and the projection of a spherical surface on a flat image also give rise to the geometric errors. These deformations, if not properly accounted for, will prevent meaningful comparison among images acquired at different times, by different sources, and with different geometries. In applications which involve change detection or image enhancement, the images must be geometrically registered to one another. In mosaicking multiple scenes, geometric corrections are also required to assure that common ground features from diverse data sets are identically positioned. In other applications which require precise geographical positioning such as cartographic mapping or temporal analysis for climatology at certain pixel locations, image navigation is needed.

Registration.

The registration procedure consists of (a) establishing a set of control points: to identify the image coordinates on the reference image corresponding to a set of well distributed image coordinates on the input image, (b) calculating the deformation model: to fit a high order-polynomial through these two sets of control points, c) resampling: to use a resampling routine on a grid generated by the polynomial to remap the input image to its reference image. In certain cases where a large set of well distributed ground control points (GCPs) cannot be easily obtained, the deformation grid can be generated by a navigation procedure. This is the case in NOAA AVHRR and METEOSAT images, and it will be discussed later. For the resampling, several methods can be used. For example, the closest neighbour selects the intensity of the closest input pixel and assigns that value to the output pixel. The bilinear interpolation uses four neighbouring input values to compute the output intensity by two-dimensional interpolation. The cubic convolution uses 16 neighbouring values to compute the output intensity (Bernstein, 1983).

Navigation.

Navigation is used to designate all operations that relate the geographic coordinates (latitude, longitude) to the image coordinates (line number, pixel number) or vice versa. One possible method is to use a complete model of the Earth, satellite, orbit, and scanner to simulate the image geometry. However, as tracking and spacecraft attitude data are not known precisely and are usually not easily available, this option is not always feasible. The compromise is to use nominal models of the Earth, satellite, orbit, and scanner along with few GCPs to adjust a number of important parameters from their nominal values in the satellite-Earth system. These calculated parameters are considered to be the effective values for the image because certain approximations have been introduced into the models and the precise values for all parameters are not obtainable. With these effective values, one can relate the geographic coordinates to the image coordinates or vice versa with acceptable accuracy. This principle has been applied to both NOAA AVHRR and METEOSAT images.

For NOAA AVHRR images, the model assumes a spherical Earth, circular orbit, and takes into account the effects due to both the Earth rotation and oblateness and the scan skew. One GCP is used along with the nodal longitude, ascending time, and the time of first scan line to calculate the effective satellite altitude and inclination angle. No other satellite information is required. The calculated altitude and inclination angle are used to identify the image coordinates from the geographic coordinates, or to locate the geographic coordinates from the image coordinates. The average root mean square (rms) errors are about 2 lines and 2 pixels (Ho and Asem, 1984). Note that the mean positioning error of the geographic grid provided by NOAA/NESS (National Environmental Satellite Service) with the image tape is ±3.7km or more than ±3 nadir pixels (Clark and LaViolette, 1981).

For METEOSAT images of the ALPEX format, the model makes use of a mathematical model of the satellite-Earth system and two GCPs to calculate the effective satellite altitude, sub-satellite point (nadir point), and the attitude yaw. The image coordinates or the geographic coordinates can then be computed from their corresponding geographic coordinates or image coordinates, respectively, by using the model with the above effective parameters. Results have shown that the average rms errors are less than 2 lines and 2 pixels for both the visible and infrared images (Ho and Asem, 1985).

The GCPs can be located automatically by using a maximum correlation search algorithm (e.g. Bernstein, 1983). The navigation procedure can be used to register one image to another or to rectify an image to certain projection systems.

One handy application of the navigation procedures is to trace the locii of pixels viewed by satellites with the same zenith angles.

Figure 9.1. Loci of pixels of the same satellite zenith angles on the NOAA and METEOSAT night infrared images of November 3, 1982 over Europe. The METEOSAT image is on the left; the NOAA image is no the right. They are both in the raw format: METEOSAT North-South axis is upward and NOAA North-South axis downward.

Figure 9.1 illustrates the result of this procedure applied to the NOAA and METEOSAT images of November 3, 1982 over Europe. These pixels are of interest because their signals received by the satellites are supposed to propagate through similar atmospheric paths and thus suffer from similar atmospheric effects (Beriot *et al.*, 1982).

Let i and j denote line and column of the NOAA image, N denote NOAA and M denote METEOSAT, F and L denote first and last respectively, and z for zenith angle, the procedure can be summarized as followed:

for $j = 1, 2, 3, \ldots, n$ **do**
    calculate $z^N(j)$;
    calculate geographic coordinates of pixels (i_F, j) and (i_L, j)
        and their corresponding z_L^M and z_F^M;
    if $z_F^M \leq z^N(j) \leq z_L^M$ **then** binary search for $z^M = z^N(j)$
    ($z_F^M < z_L^M$ by definition);
end

This procedure is a must for the atmospheric correction method proposed by (Holyer, 1984) and can also be used in the intercalibration of NOAA and METEOSAT data.

Radiometric Corrections.

The errors in image radiometry come from internal and external sources. The internal sources may be the detector response (like banding and stripes of LANDSAT) or calibration source error (like the blackbody reference of METEOSAT). The external sources are mainly the emission, scattering, and absorption of the atmosphere.

Calibration.

Each satellite has a different calibration procedure. For LANDSAT MSS for example, the radiometric calibration can be done by using the general algorithm of (USGS, 1979). For NOAA AVHRR, the calibration coefficients are given with the image tapes. For METEOSAT, the conversion coefficients for infrared data are issued periodically by ESOC (European Space Operations Center, Darmstadt, Germany). The METEOSAT visible data can be converted into radiance by using coefficients given in literature (e.g. Kriebel, 1983). In addition to the absolute calibration, relative calibrations are sometimes also necessary, particularly for LANDSAT data in case of stripes, banding, or for mosaicking the scenes. This supplementary calibration does not attempt to make the detector outputs correct. It attempts to balance their intensities.

Atmospheric Correction.

The atmosphere introduces scattering, absorption and emission effects to the signals received by the satellite. Depending on the wavelength ranges, different techniques have to be used to account for these noises.

Visible Range.

The important effect on this wavelength range is the atmospheric scattering or haze. The first-order correction is to establish the offset grey level for each spectral band. The offset corresponds to the amount of haze in the bands and is subtracted from each pixel value. One may also use the spectral band which is least affected by the haze (like band 7 of LANDSAT MSS) to correct the other bands (Sabins, 1978).

Infrared and Near-Infrared.

The main effect on this range is water vapor absorption. There is a number of techniques to correct this effect. They can be classified into two categories: multi-angular and multi-spectral methods. These techniques take advantage of the differential absorption of different spectral windows and viewing paths to eliminate the atmospheric effects (e.g. McMillin, 1975; Chedin *et al.*, 1982; Strong and McClain, 1984)

9.3. Data Presentation and Storage.

Map Projection.

In earth science applications, the final products are mostly thematic maps whose true terrestrial properties such as area, shapes, distances and direction are conserved. A conformal map projection is therefore needed. Map projection is also used in geocoded databanks for consistency with other geographic information systems where data are stored in a cartographic reference.

The most widely used projection in remote sensing is the Universal Transverse Mercator (UTM). This involves projecting the Earth surface on cylinders touching the Earth along its meridians. This projection is well suited for LANDSAT imagery except in the polar regions where the distortion is considerable. The Lambert conformal conic projectionmap projections sub Lambert conformal conic projection, where parallels are mapped as circular arcs and meridians as equally spaced radii, is also used.

Mosaicing.

In earth science applications, geocoded databases are needed so that homologous data points of different data sets correspond exactly to the same geographic coordinates. For example, the geocoded database used in our geology application covers an area of 430km×290km and a mosaic of the LANDSAT image is needed. Five LANDSAT scenes (212–29,212–

30,211–29,211–30 and 210–30) were used. The images were registered to the Lambert-III conic projection with a pixel size of about 100m. A number of 1/25,000 topographic maps in Lambert-III projection were used to define ground control points. Each LANDSAT scene was registered to its corresponding maps. These registered images were cut into rectangular blocks (with boundaries in the overlapping regions), assembled and corrected radiometrically. This proved to be difficult due to the limited choice of good quality images taken at the same date. The first step was to calculate the mean and standard deviation for each overlapping region between two scenes. One scene was arbitrarily chosen as the reference, and the other scene was adjusted to match the mean and standard deviation by applying a gain and bias factor. The corrected bands (MSS4, MSS5 and MSS7) of the mosaic were then displayed. Further gain and bias adjustments were found by trial and error to reduce colour boundary and to match the overall colour balance between frames. There existed, after this correction, noticeable changes in the colour at the boundaries. This was reduced by adjusting a ramp function over a number of pixels on either side of the boundary. This technique gave a satisfactory image of the Southeast France used in our geology study (see Colour Plates: Figure 9.2).

Data Storage and Coding.

Efficient encoding is necessary for economical storage and transmission of digital image data. These data can be compressed by exploiting the redundancy of the information contained or the level of information degradation accepted by the users. Encoding methods used for 2D image data can be grouped into:
- Transform coding by Karhunen-Loeve, Fourier, Hadamard,
- Cosine transformation (Rosenfeld and Kak, 1982)
- Predictive coding (Berstein, 1978)
- Run length coding (RLC) (Shu *et al.*, 1983)
- Quad-tree coding (Tamminen, 1984).

In this section, we present the RLC method for the storage and processing of multi-source data in a geocoded database. Emphasis is placed on the direct processing of data stored under RLC without converting back to raster format, thus reducing the processing time (Loodts and Nguyen, 1985).

Description of the implemented RLC.

The compression of a raster image into RLC is done by counting and representing identical neighbouring pixels by segments on each line. The RLC parameters can be stored under many forms, depending on the application or hardware constraints. For example, we may store the distance (or length) and value of each segment in a single file (or storage buffer), or we may store the distance (or length) on one file, and the values on another. We have chosen the second solution and the record length of the files is determined by the maximum number of segments per line. Since the data is coded in 8 bits per position, the maximum length of the segment is limited to 254 (255 is reserved to indicate the end of each line). The value of the segment is coded in an arithmetic base of 255, hence the range of values is given by 255 raised to the power of N (N is the number of files used).

Efficiency of RLC in coding Different Data.

A test was performed on different data and two important parameters were computed as followed:

$$\text{Maximum compaction ratio} = \frac{2 \times \text{total number of segments}}{\text{total number of pixels in the images}} \qquad (9.1)$$

$$\text{Implemented compaction ratio} = \frac{2 \times \text{maximum number of segments per line}}{\text{number of pixels per line}} \qquad (9.2)$$

The maximum compaction ratio indicates the compaction achieved when variable record length files are used, but in reality it is more convenient to use the fixed record length files, hence the efficiency of RLC is shown by the implemented compaction ratio.

From the results shown in table 9.1 and figure 9.3, we can find that it is only effective to store the processed data where there exists important amount of information redundancy.

Table 9.1. Compression by **R**un **L**ength **C**oding of different data

Type of Image	Max No of Segments	Compaction Ratio
LANDSAT (MSS6)	1024	0.988
Original HCMM	797	0.799
HCMM (classified)	330	0.332
Digitized map	103	0.215
HCMM (segmented)	164	0.161
Gravimetric map	142	0.115
HCMM (contour)	122	0.095

9.4. Applications in Geology.

Satellite data give a repetitive, synoptic view of a region in the visible, near infrared, and thermal infrared spectrum. Information on important structures is available in geophysical data and elevation data. These multi-source data, in digital form or digital convertible forms, allow geologists to apply recent image processing techniques to extract structural and lithologic information. The overall objective of the geology project is to develop computer-assisted geologic mapping capabilities. To achieve this, several subjects are being studied:
- Creation of a geocoded database with the following data:
    — A mosaic of LANDSAT MSS data,
    — HCMM data,
    — NOAA data,
    — Gravimetric measure,
    — A digital terrain model,
- Discrimination of lithologic materials by their spectral signatures,
- Development of new image processing techniques for analyzing geologic structures,
- Improvement and creation of regional geologic maps.

The use of LANDSAT/HCMM Data for Lithographic Analysis.

The general aim of this section is to present a study of the geologic utility of LANDSAT and HCMM data. These data are registered and used for the discrimination of various geologic formations (Abrams *et al.*, 1985).

Description of data and region studied.

The data utilized consist of LANDSAT MSS data (scene 210-30 of 16 June, 1980), and HCMM visible and thermal data (31 October, 1978). The HCMM satellite in operation between 1978 and 1980, provided images in two spectral bands: the visible and near infrared (0.5-1.1μm), and the thermal infrared (10.5-12.5μm). Data were collected during the day and at night, with a ground resolution of 500m (Price, 1978). In addition to the visible and thermal images, a derived image was available, called Apparent Thermal Inertia (ATI). This image is calculated using the formula:

$$\text{ATI} = k \times \frac{1 - \text{albedo}}{T_{\text{day}} - T_{\text{night}}} \quad (9.3)$$

where T_{day} and T_{night} are the registered day and night temperature images, albedo is calculated from the visible and near infrared image, and k is a scaling factor.

The area studied is the western Provence region of southeastern France. It presents numerous outcrops of light coloured rocks, essentially sandstones, limestones, and dolomites of various ages. They are often difficult to separate even in the field. Vegetation cover varies from 100% pine forest cover at higher elevations, to a more modest cover of 10 to 20% at lower elevations. Valleys are typically cultivated or used for cattle grazing.

LANDSAT MSS data allow separation of materials whose reflectances differ in the visible and near infrared wavelength regions. Even though their spectral responses are significantly different when measured using broken, fresh surfaces, there is little difference between the responses of their natural, weathered surfaces. The visible and near infrared reflectances are strongly influenced by the presence of lichen and the development of surface weathering rinds. On the other hand, the thermal characteristics of limestones and dolomites are distinctly different. Sabins, 1978 reports thermal inertia values for dolomite of 3138 and for limestone of 1883 ws$^{0.5}$ m$^{-2}K^{-1}$ due to large differences of diffusivity and conductivity (thermal inertia is $\sqrt{K\rho c}$, where ρ is the density, K is specific heat, and c is conductivity). This large difference should produce differences in the surface temperature. For the region studied, therefore, thermal data should be useful in the discrimination of limestone/dolomite terrains.

Data Registration.

The data from LANDSAT and HCMM were registered by the method mentioned previously. Due to the great disparity in pixel size (80m versus 500m), both data sets were resampled temporarily to 250m to facilitate the acquisition of ground control points. The two data types were displayed side by side on a display console, and common points were identified on each. The deformation model to transform the HCMM data to the LANDSAT geometry was a fifth degree polynomial surface which best fitted the control points. Finally, the HCMM data were transformed using this model and resampled to 80m pixels (Figure 9.4a).

Visual Display and Interpretation.

There are many techniques available for displaying multivariate image data; the most commonly used is colour compositing using red, green and blue. The LANDSAT colour composite (Figure 9.4b) was produced in this manner, projecting MSS channels 4, 5, and 7 in blue, green, and red, respectively. The composite LANDSAT-HCMM image (Figure 9.4c) was produced in a different manner, using the colour space Intensity, Hue and Saturation. The intensity channel was the first principal component computed from the LANDSAT channels, the hue information came from the HCMM night-time thermal image. The saturation was kept constant.

Various visualizations were analyzed using the ATI image. However, the ATI image was less useful than the night-time thermal image. The reason seems to lie in the nature of the ATI image. Because the range and magnitude of the night-time temperatures are much less than those of the day-time temperatures, the difference is not very different than the day-time image minus a constant. The day-time data are strongly affected by topography and insolation anisotropies, and hence contain very little geologic information, making the ATI images of little use in this region.

In certain areas, the LANDSAT data allow the recognition of formational boundaries, and there are sufficient contours discernible for satisfactory mapping. In other regions, however, the boundaries visible on the image are not sufficiently distinct to draw an interpretive map with great confidence. In addition, there are areas where no geologic boundaries can be detected. Two such regions were selected for analysis using the LANDSAT-HCMM composite, to evaluate the additional geologic information presented by combining reflectance and emittance data.

The first region (A) is located between St.-Maximin-la-Ste.-Baume and the syncline of Rians; outcrops consist of Jurassic to lower Cretaceous calcareous rocks. The second region (B) is located west of Cuers, and is underlaid by Triassic to lower Jurassic formations. In area A the formations appear white to light yellow on the LANDSAT image and cannot be differentiated. On the combined image, it is possible to separate the Jurassic formations which are dominantly limestones from those which are marno-calcareous. In addition, a zone of dolomitization in the Cretaceous formations appears on the image. In area B two synclines are found, with dolomites stratigraphically above argillites. In the northern syncline, the dolomitized Jurassic rocks are clearly distinguishable from the undolomitized limestones. In the southern syncline, the contact between the Triassic marnocalcareous rocks and the Jurassic dolomites is visible. None of these contacts are discernible on the LANDSAT composite alone.

Results.

This example demonstrates a procedure for utilizing two image data sets with greatly different spectral and spatial characteristics. It is feasible to preserve the higher spatial resolution data and to resample the low spatial resolution data to the same pixel size. The IHS, display using the full resolution LANDSAT data as intensity and the HCMM as colour provides a good image for photo-interpretation. The geologic importance of combining data from different spectral ranges is evident from the added discrimination of carbonate terrains.

The detection of linear geological structures.

Photo-interpretation techniques allow geologists to identify geological structures on images. In this section we are interested in the detection of lineaments in satellite and gravimetric data. Lineaments, straight or gently curved, lengthy features on the Earth surface, can be identified (Fraipont and Hirsch, 1984) by the linear shape of:
- the contact between two lithological units with different spectral properties,
  - a change in the vegetation cover,
  - an anomaly in the vegetation cover,
  - the edge of a water basin,
  - a river network,
  - a topographic relief.

These linear features appear on the images as: linear boundaries between two zones of different brightness or different textures, linear shadows, linear inclusions.

A few algorithms for the detection of linear structures on LANDSAT, HCMM and gravimetric data will be presented.

The detection of lineaments on LANDSAT data.

The extraction of contours having a strong local gradient was accomplished by using an algorithm developed by Asfar (1981). The process consists of:
1. Detection of edges using a generalized Sobel operator,
2. Selection of points having a local maximum gradient,
3. Search for the nearest neighbours of each point, taking into account the gradient direction,
4. Construction of the contours from the neighbour image.

For each detected contour (figure 9.5a) several shape parameters can be calculated:

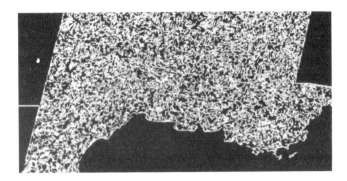

Figure 9.5a. Detected contour on LANDSAT data

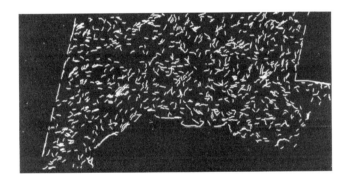

Figure 9.5b. Two dimensional histogram and the selected linear contours on LANDSAT data

Length: the contour length is calculated by following the contour; horizontal and vertical increments are counted as 1 and diagonal increment as $\sqrt{2}$.

Linearity: if the distance between the two extremities of each contour is D and the length is L, the index of "nonlinearity" can be defined as $R = L/D$. The value of R is 1 for linear contours, and increases with more sinuous contours.

Direction: the contour direction is the direction of the straight line passing through the contour extremities.

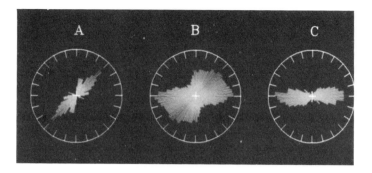

Figure 9.5c. Rose diagrams showing principal direction of lineaments on LANDSAT data

Lineaments as desired by geologists can be selected by defining clusters in a 2 dimensional histogram representing the length and the non-linearity index of the contours (Nguyen *et al.*, 1984a) (figure 9.5b). Once the linear contours are detected, the contour image can be used to mask the angle image produced by the gradient program. The resulting masked image is then used to generate the rose diagrams (Nguyen *et al.*, 1984b). These rose diagrams show the principal direction of the lineaments (figure 9.5c)

The detection of elongated high gradient zone on HCMM data.

The HCMM night thermal infrared (10.5-$12.5 \mu m$) data (Figure 9.6a) are used to discriminate the different surface materials. The visual interpretation of the data reveals that the lithologic units are heterogeneous on a scale of a few pixels. The heterogeneity is due to local disturbance caused by the shadowing, the vegetation cover, the land use. In order to homogenize the image, a segmentation procedure is applied to the data using a contour extraction program (Asfar, 1981). The segmented HCMM data is coded in RLC and the compaction ration is 0.161 compared with 0.799 for the original data. The geologists are interested in "thermal linears" defined as linear anomalies located in a cold or warm zone and linear frontiers between cold and warm zones. In the segmented data, these can be characterized by segments with an elongated and linear shape.

Calculation of perimeter: the perimeter is defined as the number of pixels belonging to the border of each surface. This is done by accumulating the number of pixels of the same surface not surrounded by pixels belonging to the same surface in two consecutive lines (Figure 9.6b).

Calculation of area: this is done by accumulating the length values of

Figure 9.6a. HCMM night temperature data

Figure 9.6b. Image of the perimeter of different segments on HCMM data

pixels belonging to the same surface in two consecutive lines (figure 9.6c).

Elongateness: a procedure has been developed to give a measure of the elongateness by the parameter $P/\sqrt{A}$, where P is the perimeter and A the area of each segment. Elongated segments can be identified by their high value in the ratio $P/\sqrt{A}$ (figure 9.6d).

Detection of high gradient area on gravimetric data.

The gravimetric measure (Bouger's anomalies) contour map is digitized and converted into a binary raster image (Figure 9.7a).

The raster image is stored in RLC files with a compaction ratio of 0.115. The gravimetric data is useful in detecting major geologic structures with particular patterns, for example, linear limit or inclusion. It is then interesting to derive information concerning the gradient of the gravimetric

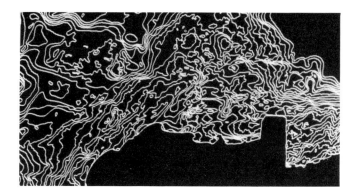

Figure 9.7a. Contour image of the gravimetric data

measure. The gradient can be calculated from the raster image derived from the contour map by interpolating (Leberl *et al.*, 1980) or by using the two-dimensional Fourier transform (Papo and Gelbman, 1984). We are proposing a simple technique of expansion/erosion of the contour lines. Since the increment, say i, of the values represented by two consecutive contours is constant, the gradient in area merged by the expansion and erosion with a circle of diameter d will be inversely proportional to d. By repeating the process with a range of values for the diameter (3,5,7), we will obtain a map of gradient (with gradient of i/3,i/5,i/7) (figure 9.7b).

Results.

The technique developed to produce automatically lineament maps from image data eliminates the subjective bias inherent in manual image interpretation. The image can be used for structural analysis, or the information can be summarized using the automatic rose diagram program. The rose diagrams produced automatically from the image are found to be in very good agreement with those produced manually and also with the microtectonic measurements made in-situ (Blusson *et al.*, 1984). The detection of linear features using three different data-sets allowed the geologist to confirm the existence of major fault systems at different depths and hence to provide a more complete, 3-dimensional picture of the geologic environment.

Exploitation of LANDSAT/DTM data.

Topographic data.

For geologists, the topographic features of the terrain are important both for the detection of geologic formations and the study of tectonic movements. Normally this information is obtained from topographic maps or stereo air photographs. LANDSAT data, due to the very small base-to-height ratio, have very poor stereo capability. At mid-latitudes, the overlap between adjacent orbits is also minimal. Therefore, to obtain topographic information from the LANDSAT data it is necessary to use ancillary elevation data. In this study a digital elevation model (DEM) was used, giving the altitude at each point of a regular grid of 50m interval (figure 9.8a).

The region studied is the Mont Saint Victoire; the data has 319 lines and 419 pixels covering an area of 15.6 by 21 km. The DTM is used to generate a synthetic reflectance image. It can also be used together with the LANDSAT data over the same area to: generate 3-D views of the LANDSAT data; emphasise the relief by modulating the LANDSAT data with the synthetic reflectance image; create a stereo pair of LANDSAT images.

The generation of synthetic reflectance image.

The reflectance of a uniform surface is a function of the position of the Sun, the local slope of the surface and the observer's position. Using The Lambertian law of reflectance and a vertical observer's angle, the reflectance (R) at a point on an image is given by:

$$R = \max(0, k \cos \alpha) \qquad (9.4)$$

where k is a coding coefficient (equal to 255 in our case) and α is the angle between the normal to the surface and the Sun direction. The Sun angle and elevation of the corresponding LANDSAT image is used in the generation of the synthetic reflectance image as shown in Figure 9.8b.

The generation of LANDSAT 3-D Views.

The LANDSAT data are first registered to superimpose exactly on the DEM. Using the three LANDSAT bands as primary input colours (red, green and blue) and the DEM to calculate the position of a point viewed from a direction (azimuth and elevation), a 3-D view is generated, as shown in Figure 9.8e.

The generation of LANDSAT stereo pairs.

A variation of the previous method allows the generation of a stereo pair of images by using the same elevation angle and complementary azimuth angles. A simple rotation will produce a pair of LANDSAT images for stereo viewing, as shown in figure 9.8f.

Results.

The LANDSAT image modulated with the DEM (figure 9.8d), compared with the original LANDSAT image (figure 9.8c), already shows a dramatic improvement in the display of the topographic relief. The original data were acquired in June, when the Sun's elevation was relatively high, thus suppressing shadowing and perception of relief. Addition of the DEM data allows an interpreter to evaluate the relationship between lithologic composition (colour) and morphology. Folds are easily seen, both by their semi-circular outcrop patterns and by their topographic appearance as curving mountain ranges. It is even possible to estimate the dip direction and magnitude from this image. These parameters are very useful in making a more complete geologic map. The 3D perspective views (figure 9.8e) allow the interpreter to change his viewing perspective of the landscape to have an optimal display of particular structural features. For example, the view with azimuth of 160 degrees shows more clearly a linear feature expressed as an elongated ridge. This feature is a mapped fault unrecognizable on the LANDSAT data alone due to its lack of distinctive spectral contrast. The stereo pair (figure 9.8f) is a tool familiar to all geologists, who are trained to use stereo air photographs in image analysis. Again, this display provides the missing third dimension of information.

9.5. *Applications in climatology.*

It has been recognized that the global environment has undergone constant natural and man-induced changes. A worldwide coordinated effort has been proposed to investigate these changes and contribute to our still rudimentary knowledge in this aspect. This effort involves many interdisciplinary studies using land, sea, and space-based data. Particularly, many satellites are currently providing information on the state of the Earth's surface such as albedo, temperature, and cloud. Unfortunately, many satellite data are undersampled or of different degrees of precision. In order to take full advantage of the specific characteristics of each satellite system, we have to incorporate many different sources to infer meaningful parameters at the Earth's surface. These parameters serve in monitoring seasonal and long-term variations in land use, vegetation cover, surface structure and characteristics. These parameters can also be used to derive other 'secondary' parameters such as thermal inertia, soil fluxes, and soil moisture which are essential in climate studies and agricultures. In fact, these parameters can be also measured at the meteorological stations around the world. However, the stations are sparsely distributed, and their measurements are not uniform. Only satellite data can provide spatial regularity and measurement uniformity for these parameters.

In this section, we will illustrate a number of applications of satellite imagery to Climatology problems. The data used are mainly from METEOSAT due to its high temporal coverage rate (every half hour). However, NOAA AVHRR data are also used to recalibrate METEOSAT data because of its good radiometric calibration and spatial resolution (1 km × 1 km at nadir). We will discuss the extraction of primary information such as surface albedo and temperature followed by the extraction of secondary information such as soil fluxes and thermal inertia.

Extraction of primary information.

Albedo.

This information can be obtained by dividing the radiance derived from the visible METEOSAT data by the solar irradiance and the cosine of the local solar zenith angle. The geographic location of the pixel is determined by the navigation procedure for METEOSAT images (Ho and Asem, 1985).

Temperature.

The temperature can be derived from the infrared data. However, there is no proper procedure for correcting the atmospheric effects on the METEOSAT temperature data; we have used the corrected AVHRR data to calibrate the top-of-the-atmosphere METEOSAT data by the regression method (Ho, 1985). The atmospheric effects on AVHRR data are eliminated by using the split window correction procedure. The correction coefficients for NOAA data have been generated locally by using ship-measured sea surface temperatures in the area and their corresponding radiometric temperatures on NOAA. The ship locations have been located on the image by our navigation procedure. The correction results are similar to those obtained by using the coefficients given by Strong and McClain (1984).

Figure 9.9 illustrates the result of all the corrections discussed above. The registration of the temperature images derived from NOAA and METEOSAT infrared data of November 4, 1982 is shown above. The NOAA temperatures have been corrected for the atmospheric effects. The METEOSAT image is the top-of-the-atmosphere temperature. They are both rectified over a rectangular geographic grid covering Tunisia. The two-dimensional histogram over the desert area south of Tunisia is shown below (see the rectangle on the image). A highly linear correlation can be easily observed. A linear regression has been done for this zone to establish the correction coefficients to eliminate the atmospheric effects on METEOSAT temperature data (gain: 0.912, bias: 4.445, sample correlation coeff.: 0.90). A similar operation has been also done for the night images of NOAA and METEOSAT to provide the correction for low temperature values (gain: 1.048, bias 2.146, sample correlation coeff.: 0.84). This approach has implicitly assumed the atmospheric effects on the soil are similar to those on the sea and has ignored also the time dependent atmospheric effects on the infrared data. However, as the correlations are high, we expect the induced errors to be negligible. These atmospherically corrected temperatures and calculated albedoes are used in the climatological model discussed below.

Extraction of secondary information.

The objective is to use the albedo and temperature data of a pixel throughout the day to derive secondary parameters such as soil fluxes and thermal inertia. A number of works have been published on this subject; however, all of them require measurements of meteorological or geologic parameters at the site or at the nearby meteorological stations (see, for example, Kahle, 1977; Raffy and Becker, 1985). In this section, we will present a technique to generate the map of the combined sensible and latent heat fluxes and to calculate both the daily soil conducting flux cycle and thermal inertia by using only remotely sensed data.

Principle.

The principle is based on the energy exchange process at the soil surface level (Ho, 1985).

$$G = P \sum_{k=1} \sqrt{\frac{\omega_k}{2}} (\frac{T_k + 1/\omega_k \delta T_k}{\delta t}) = S + R + H + LE \qquad (9.5)$$

where
G: soil surface conducting flux
P: soil thermal inertia,
T_k: the kth Fourier component of the daily surface temperature curve derived from METEOSAT infrared data
ω_k: the kth Fourier component of the angular frequency
S: insolation determined by the satellite albedo data
R: thermal radiation modelled by the thermal infrared data
H: sensible heat flux, a function of local wind speed and soil-air characteristics

LE: evapotranspiration flux, a combined effect of evaporation and transpiration, a complex function of many atmospheric and soil variables such as pressure, relative humidity, and surface characteristics...

The soil conducting flux can therefore be measured by remotely sensed means provided that the soil thermal inertia is known.

Combined sensible and evaporation fluxes.

From the above equation, given the daily temperature cycle, we are able to determine the time when the conducting flux is equal to zero. At those instants, the combined sensible and evapotranspiration fluxes equals the net radiation $(S+R)$, thus derivable from the satellite visible and infrared data. Figure 9.10 shows an example of the combined sensible and evapotranspiration fluxes obtained from METEOSAT data over Tunisia on November 4, 1982 at 15:50 LT (local time). The conducting fluxes are found to vanish about 15:40-16:00 LT throughout the area (Ho, 1985). The dark tone is for low fluxes and light tone for high fluxes. Note that the area north of Chott Djerid is covered by clouds.

Figure 9.10. Sensible heat

Thermal Inertia and the surface conducting flux.

In stable meteorological conditions, it has been observed that early in the morning, at the turning point of the daily temperature cycle, say t_0, the combined sensible and evapotranspiration fluxes usually vanish. The soil thermal inertia thus can be calculated directly from the net radiation at that instant,

$$P = (S + R)/\{\sum_{k=1} \sqrt{\omega_k/2}(T_k + 1/\omega_k \frac{\delta T_k}{\delta t})\} \quad (9.6)$$

Calculation of thermal inertia has been done for a test zone south of Chott Djerid (the rectangle on Fig. 10). The values obtained are mostly from 1600–3200 w $s^{0.5}m^{-2}K^{-1}$ and within the reasonable range for the desert surface (Ho, 1985).

With the value of thermal inertia, the surface conducting flux can be easily calculated for the rest of the day by using the formula given above.

9.6. Conclusion.

The availability of satellite data in various spectral bands and at different spatial resolutions presents a formidable challenge to users to exploit these data efficiently. We have presented a few applications in geology and climatology; the results are very encouraging. The experience has also allowed us to develop a series of image processing techniques necessary to acquire and integrate data of various sources in a geocoded database, to extract useful information, to manipulate the processed data interactively, and finally to display the results. The amount of effort in the acquisition and pre-processing of the heterogeneous data was considerable; this shows the need for standardization of the data format and the interface between different geographical, geophysical information systems. More effort should be also undertaken to deal with physical modelling and decision-making process in order to equip us with more precise knowledge toward creating a geoscience expert system. For now, our best hope is a computer-aided system for geophysical information extraction.

Acknowledgements.

This chapter presents a number of results obtained from the IBM-CNRS-CNES joint study on the use of multiple source data processing in Geology and Climatology. The authors wish to thank the following members of the Paris Scientific Center for their valuable contributions: A. Asem, A. Blusson, V. Carrere, J. Y. Garinet and Y. Rabu.

References.

Abrams M., A. Blusson, V. Carrere, P.T: Nguyen and Y Rabu, 1985, Image processing applications for geologic mapping, *IBM J. Res. Develop.*, Vol 29 No 2, ,177–186.

Asfar, L., 1981, A method for contour detection, segmentation, and classification of Landsat images, *International Geoscience and Remote Sensing Symposium*, Washington, D.C., 298–304, June.

Beriot, N., Scott N.A., Chedrin A., and Sitbon P., 1982, Calibration of Geostationary-Satellite Infrared Radiometers Using the TIROS-N Vertical Sounder: Application to METEOSAT-1, *J. Appl. Meteor.*, 21, 84–89.

Bernstein R. (ed.), 1978, *Digital image processing for remote sensing*, IEEE press, NewYork.

Bernstein, R., 1983, *Image Geometry and Rectification, in Manual of Remote Sensing* (ed. R.N. Colwell), 873–922.

Blusson A, M. Abrams P.T. Nguyen, and P. Masson, 1984, Structural analysis of the Cevennes (France) using multiple data types, *International Symposium on Remote Sensing of Environment: Remote Sensing for Geological Exploration.*

Chedin, A., N.A. Scott, and A. Berroir, 1982, A single-channel, double viewing method for sea surface temperature determination from METEOSAT and TIROS-N radiometric measurements, *J. Appl. Meteor.*, 21, 613–618.

Clark, J.R. and P.E. LaViolette, 1981, Detecting the movement of oceanic fronts using registered Tiros-N imagery, *Geopgys. Res. Lett.*, 8, 229–232.

Fraipont P and J. Hirsch, 1984, Analyse lineamentaire: procedure de traitement de donnees teledetectees, *Colloque Int. on Computers in Earth Sciences for Natural Resource Characterization*, Nancy, April.

Ho, D. and A. Asem, 1984, Automatic registration of Meteosat and AVHRR LAC images, *Proceedings of International Conference in Satellite Remote Sensing*, University of Reading, September, 81–90.

Ho, D. and A. Asem, 1985, Navigation of Meteosat Visible and Infrared images of ALPEX Format and Its Applications, paper presented at the *ESA Fifth Scientific Users' Meeting*, Rome, Italy, May.

Ho, D., 1985, Soil Thermal inertia and Sensible & Latent Heat Fluxes by Remote Sensing, paper presented at the *Fourth Thematic Conference: "Remote Sensing for Exploration Geology"*, San Francisco, April.

Holyer, R.J., 1984, A two-satellite method for measurement of sea surface temperature, *Int. J. Remote Sensing*, 5, 115–131.

Kahle, A.B., 1977, A simple thermal model of the earth's surface for geologic mapping by remote sensing, *J. Geophys. Res.*, 82, 1673-1680.

Kriebel, K.T., 1983, Results of Meteosat-2 VIS channel Calibration, *Proceedings of ESOC Fourth Meteosat Scientific Users' Meeting*, Clermond Ferrand, France, November.

Leberl, F., R. Kropatsch, and V. Lipp, 1980, Interpolation of raster heights from digitized contour lines, *XIV Congress of the International Society for Photogrammetry*, Hambourg, 726–733.

Loodts J and P.T. Nguyen, 1985, The use of Run Length Coding technique in image storage and processing—an application in remote sensing, *The 4th Scandinavian Conference on Image Analysis*, Trodheim, June.

McMillin, L.M., 1975, Estimation of sea surface temperature from two infrared window measurements with different absorption, *J. Geophys. Res.*, 80, 5113-5117.

Nguyen, P.T., S. Simon, L. Asfar, and A. Blusson, 1984a, Extraction de parametres de forme pour la recherche des structures lineaires dans les images-satellites, *Proc. First Image Colloquium*, Biarritz, May.

Nguyen P.T., S. Simon, and L. Asfar, 1984b, An application of pattern recognition for structural analysis in geology, *Int. Conf. on Pattern Recognition*, Montreal, July–August.

Papo H.B. and E. Gelbman, 1984, Digital terrain models for slopes and curvatures, *Photogrammetric engineering and remote sensing*, 50, 695–701, June.

Price, J., 1978, *Heat Capacity Mapping Mission Users' Guide*, NASA/Goddard Space Flight Center, Greenbelt, MD.

Raffy, M. and F. Becker, 1985, An inverse problem occurring in the thermal infrared bands and its solutions, to be published in *J. Geophys. Res.*.

Rosenfeld A. and A.C. Kak, 1982, *Digital Picture Processing*, Academic Press, New York.

Sabins, F.F., 1978, *Remote Sensing Principles and Interpretation*, W.H. Freeman and Co., San Francisco.

Shu D., Y.N. Sun, and C.C. Li, 1983, Run length based image segmentation, *Proceedings of IEEE Conference on Computer vision and pattern recognition*, 154–156, June.

Strong, A.E. and E.P. McClain, 1984, Improved ocean surface temperatures from space-comparisons with drifting buoys, *Bull. Amer. Meteor. Soc.*, 65, 138–142.

Tamminen M., 1984, Encoding pixel trees, *Computer vision, graphics and image processing*, 8, 44–57.

USGS, 1979, *Landsat Data Users Handbook*.

10
Extreme Variability, Scaling and Fractals in Remote Sensing: Analysis and Simulation

S. Lovejoy,
McGill University, Physics Dept.,
3600 University st.,
Montreal, Que.,

D. Schertzer,
EERM/CRMD,
Météorologie Nationale,
2 Ave. Rapp,
Paris 75007

10.1. Overview.

The extreme variability (intermittency) of remotely sensed data poses challenging new problems of sampling, averaging, calibration and modelling. New developments in non-linear dynamics, particularly in chaos, fractals and hydrodynamic turbulence suggest systematic methods for attacking these problems, involving both the stochastic modelling of scales smaller than the remote sensor resolution, as well as new data analysis techniques.

The extreme variability discussed here has two distinct aspects: firstly, fluctuations span a wide range of scales and secondly, even at a given scale, they may span a wide range of intensities. This large variability of geophysical phenomena leads to statistical properties quite different from those usually assumed. The first aspect leads to a strong scale dependence of the statistical properties of the space (and/or time) averaged quantities. In particular, as we average over larger scales, low powers of the field increase, while those of high powers decrease. In between, there is physically significant power which has scale invariant averages. The second aspect is associated with a sensitive dependence on the dimension (e.g. line, plane, volume) over which the averages are taken. The scale dependence of the averages occurs because the scales of interest are smaller than the characteristic outer scale of the variability, which for many geophysical quantities is likely to be of the order of thousands of kilometers. While this scale dependence is widely recognised (at least implicitly), it has not been adequately investigated, and the study of the rainfield reported here is probably the first of its kind.

The second dependency noted above—the dimensional dependence of averages—is a new possibility arising out of the recent discovery of mathematical measures with multiple fractal dimensions. Again, radar data is used to measure the scale dependence of the rainfield at 1, 1.5, 2, 3, 4 dimensions (space and time), clearly showing the sensitivity of averages to the dimension over which they are taken. Empirically, we find that it is the radar reflectivity rather than the rain rate which is scale invariant. The dimensional dependence of the averages is important because we also show, by analysing the Canadian ground station network, that at least over the range of $\approx$ 30km to 1000km, that it has a fractal dimension of about 1.5, reflecting the fact that the stations are distributed in an extremely inhomogeneous fashion, primarily being concentrated near population centers.

10.2. Introduction.

In the past twenty years, the acquisition, transmission, and analysis of remotely sensed information has been gradually transformed from a highly specialised technique primarily of interest to cartographers, into an operational field of industrial proportions. The quality, quantity and diversity of remotely sensed information, as well as our ability to process it, is evolving so quickly that in many areas users have not been able to keep up. For example, notably in meteorology, a gap has developed between the enormous quantity of data routinely available, and the tiny fraction actually used in key areas such as numerical weather prediction. The cause of this significant under-exploitation of existing information lies in both institutional and human inertia, as well as in the real (scientific) difficulty of the subject itself. On the other hand, remote sensing devices detect fields which are seldom quantities of immediate interest in themselves and, furthermore, are often extremely variable over a wide range of time and space scales.

The first of these difficulties is aggravated by the fact that remotely sensed information is often exploited according to long established concepts and procedures. The real physical significance—and even certain essential empirical characteristics—of the data are often ignored, and the data mutilated (for example for use in parametrisation schemes). Alternatively, they are only used in a qualitative and very indirect manner, for example the use of cloud pictures to estimate winds, or to adjust the position of meteorological fronts.

The second difficulty is the primary subject of this paper: extreme variability. The problems inherent in dealing with highly intermittent, discontinuous fields, are exaggerated by the fact that remotely sensed data are usually analysed with standard statistical techniques involving the assumption both of the existence of characteristic time and distance scales (hence an exponential fall-off of correlations), and the existence of characteristic amplitudes of fluctuations (hence an exponential fall-off of probability distributions). However, it has been known for some time, especially in the field of hydrodynamic turbulence, that these methods are ill-suited for dealing with intermittent phenomena (involving the sudden appearance or disappearance of structures, sharp gradients, etc). Hence, the common complaint that the statistics are non-stationary, are too variable, or both. Conversely, a variety of methods (in particular, those based on scale invariance and intermittency) have been developed in the context of turbulence and more generally, of dynamical systems, but have only very recently been applied to the analysis and simulation of remotely sensed data.

Basing ourselves largely on the example of the atmosphere, we argue that the type of extreme variability relevant to remote sensing can be understood in terms of a symmetry principle that we call Generalised Scale Invariance (GSI—Schertzer and Lovejoy, 1985a). GSI grew out of developments in hydrodynamic modelling in which the variability was considered to arise as a result of a cascade process concentrating energy from one scale to another, while conserving its average value (see e.g. the simple cascade scheme shown in figure 10.1).

Extreme variability results because as the number of cascade steps tends to infinity, the energy becomes distributed over a (mathematically) singular measure which may be characterised by a hierarchy of fractal dimensions. That fractals are involved should come as no surprise, since the cascade we have just described involves no characteristic length and is therefore scale invariant (abbreviated "*scaling*"). However, what makes the variability here different from (and much stronger than) the usual fractal models, (such as the fractal mountains shown in Voss (1983), Mandelbrot (1982), and Fournier *et al.* (1982), is that the the most intense regions are characterised by lower fractal dimensions than those of weaker regions. Specifically, the fractal measure ϵ (i.e. energy flux) associates a "mass" with each point in space. On this measure, the threshold ϵ_T defines a fractal set with dimension $D(\epsilon_T)$. The requirement that the intense regions are subsets of the weak regions implies that D is a decreasing function of ϵ_T. For a region size L, the number of points on a fractal set varies with L^D, hence sets with small D (e.g. intense regions) are very sparsely distributed (see Hentschel and Proccacia, 1983, Grassberger, 1983, Schertzer and Lovejoy, 1983 Parisi and Frisch, 1985, Mandelbrot, 1984a and Schertzer and Lovejoy, 1985a,b).

GSI is a formalism that develops these cascade ideas in two directions. First, it shows with the aid of a generalised scale changing operator, how

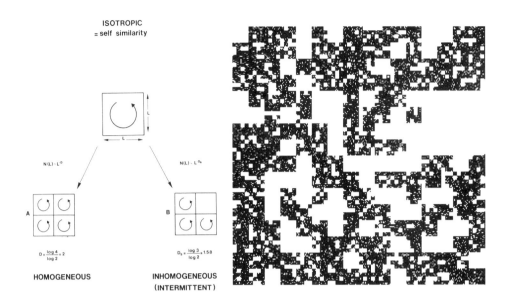

Figure 10.1a. (a) schematic representation of how various turbulence schemes treat the break-up of a single eddy (represented by the central square) via non-linear interactions during a single step in the cascade process. Both schemes shown here are isotropic, the left hand side is homogeneous, and the right hand side, intermittent. (b) Here we show the full cascade obtained by iterating eddy breakup a large number of times. The black regions are active, the white, dead (for more details of this β model, see Section 10.6). On the fractal (black, dimension D_S), the number of active eddies (N) at a given size L varies as $N(L) \propto L^{D_S}$.

anisotropic scale invariance (for example that between the horizontal and vertical directions in the atmosphere) can be accounted for and modelled. Second, it leads to a systematic study of the various measures that can be scale invariant under the action of the scale changing operators. These scale invariant quantities are expected to have a fundamental physical significance, since the ensemble average of their spatial means does not depend on the scale (or dimension of space) over which they are averaged. Conversely, quantities defined by the various powers of the scale invariant quantity are strongly scale dependent. Precisely, the moments $\langle \overline{\epsilon}^h \rangle$ (where $\overline{\times}$ indicates spatial, and $\langle . \rangle$ indicates statistical, ensemble averages) increase with scale for $h < 1$, and decrease for $h > 1$.

In hydrodynamics, it is plausible that flows respect GSI because the governing (Navier-Stokes) equations contain no length scale above a small inner viscous scale, which in the atmosphere is of the order of millimeters.

Hence, scale invariant solutions such as those displaying a Kolmogorov $k^{-5/3}$ energy spectrum are possible. However, in remote sensing, the exact equations governing the phenomena of interest are seldom known, hence we have recourse only to more general arguments of the type that physicists call symmetry principles.

Familiar symmetry principles include the conservation of energy, and matter as well as invariance with respect to translation, rotation and reflection. Following a standard approach in physics, a system is expected to respect symmetries unless specific ("symmetry breaking") mechanisms can be found. In any case, the widespread existence of power-law (and hence scale invariant) spectra in geophysical quantities (e.g. Deschamps et al., 1981, Mitchell, 1982, Lilly and Peterson, 1983, Nastrom and Gage, 1983, Muller, 1985a,b, Lovejoy and Schertzer, 1985b), as well as other evidence (e.g. Mandelbrot and Wallis, 1969, Burroughs, 1981, Lovejoy, 1981,1982, Rothrock and Thorndike, 1983, Kagan, 1981a,b, Kagan and Knopoff, 1980, Korchak, 1938, Bradbury and Reichelt, 1983, Goodchild, 1982, Bills and Kobrick, 1985), shows that in many important situations, over large ranges in scale, no mechanisms exist that are strong enough to break the basic scaling symmetry. Hence, over ranges spanning a small inner scale and large outer scale, GSI is likely to be relevant. In the atmosphere, the inner scale is due to viscosity, and is of the order of millimeters. The exact size of the outer scale is not well established, but is probably at least 1000km. See Lilly (1983) for an excellent review of the now extensive empirical evidence, and Schertzer and Lovejoy (1985a) for a discussion and numerous references.

The systematic exploitation of the idea of scale invariance in remote sensing has only recently begun (see Lovejoy, 1981,1982 and Cahalan et al., 1984), but it promises to play an important role not only in calibration and sampling problems, but also in the analysis and modelling of remotely sensed information. In this chapter, we discuss each of these points giving examples mainly from the authors' own investigations and models of remotely sensed atmospheric data. In Section 10.3, we discuss and review various analytical techniques which are generally applicable to studies of the scale dependence of phenomena, giving examples using rain and cloud data. We also introduce a new analysis method called the integral structure function which is specifically designed to analyse multidimensional phenomena. In Section 10.4, we argue that stochastic models of remotely sensed phenomena are essential if their statistical properties are to be fully understood. We also show how to construct a simple mono-dimensional model that has sufficiently strong variability that it can be used to model the scale (although not dimension) dependence of rain and cloud fields. Although the mono-dimensional nature of such models is a basic limitation, their extreme variability and scaling properties combine to yield simulations with many realistic features (both visual and statistical), such as evolving cloud fields with complex structures, clustering, and texture (readers are urged to survey some of the figures before continuing). Next, we review simple cascade models and the elements of GSI, showing the limitations of mono-dimensionality. In Section 10.5, we analyse the problem of ground station (*in situ*) calibration of remotely sensed data, and show, using the example of the Canadian meteorological surface network, that its dimension (D) is ≈ 1.5 which is considerably less than that of most remotely sensed data (which are typically averaged over surfaces or volumes, hence $D = 2$ or 3). Finally, we corroborate our analysis with stochastic simulations of ground station networks. The statistical properties of multidimensional measures can depend strongly on both the dimension and scale over which they are averaged (e.g. when $h = 1$ in the preceding), hence, the differing dimensions of *in situ* and remotely sensed data pose a fundamental and challenging problem for the exploitation of remotely sensed information. Calibration is far more difficult than in the simpler mono-dimensional case and requires the solution of a number of theoretical problems. By considering high order structure functions (Section 10.6) we show how multidimensionality affects averages of various powers of a measure when the scale is changed. Using radar rainfall data, we confirm empirically that the statistical properties of the rain field are sensitively dependent, not only on the scale, but also the dimension (e.g. line, plane, volume or fractal set) over which they are averaged.

10.3. Analysis Techniques for Mono and MultiDimensional Phenomena.

Discussion.

For a function of a single variable $X(t)$, a particular type of scale invariance arises when fluctuations (ΔX) at large scale ($\lambda \Delta t, \lambda > 1$) are related to those at small scale (Δt) by the following simple relationship:

$$\Delta X(\lambda \Delta t) \stackrel{d}{=} \lambda^H \Delta X(\Delta t) \quad (10.1)$$

where $\stackrel{d}{=}$ means equality in probability distributions; that is, $X \stackrel{d}{=} Y$ if $\Pr(X > q) = \Pr(Y > q)$ for all q, where "Pr" means probability. $\Delta t = t_1 - t_0$, $\Delta X(\Delta t) = X(t_1) - X(t_0)$, $t_2 = t_0 + \lambda(t_1 - t_0)$, $\Delta X(\lambda \Delta t) = X(t_2) - X(t_1)$. H is a constant called the scaling parameter. If X is a general function of position vector $\vec{x}$ then the scale changing operation (written above as $t \to \lambda t$) is more generally written as $\vec{x} \to T_\lambda \vec{x}$ and scale invariance then becomes (Schertzer and Lovejoy, 1985a):

$$\Delta X(T_\lambda \vec{x}) \stackrel{d}{=} \lambda^H \Delta X(\vec{x}) \quad (10.2)$$

where T_λ is a generalised scale changing operator, which can involve differential rotations stratifications and (even) other more complex transformations (in addition to the basic "zoom" or magnification). For example, in linear GSI, $T_\lambda = \lambda^G$, where G is a matrix called the generator of T_λ. Under certain conditions on G, T_λ can be used to define an anisotropic metric, hence giving precise meaning to the notion of anisotropic scale invariance. Anisotropic scale changing operators were introduced in Schertzer and Lovejoy (1985a) in order to model both the vertical stratification of the atmosphere (due to gravity), as well as differential rotation (due to the Coriolis force). This is theoretically important because it establishes that the scope of scale invariance extends considerably beyond the self-similar situation where the large scale is simply a carbon copy of the small scale (e.g. Mandelbrot, 1982).

In this Chapter, we will be primarily interested in the structure of geophysical fields in the horizontal: however, consideration of differential rotation is outside our scope even though, at least in the atmosphere; it can be quite significant. We therefore simplify the discussion by restricting our attention to the special self-similar case where the operator T_λ is isotropic: $G = \mathbf{1}$, $T_\lambda = \lambda\mathbf{1}$, where $\mathbf{1}$ is the unit (identity) matrix. T_λ therefore simply magnifies structures without any deformation or rotation.

A wide variety of analysis tools exist for studying scale invariant phenomena and for determining their scaling parameters, as well as the limits of their scaling regimes. The methods we discuss have all been developed for studying phenomena that vary over a wide range of scales. They are all capable of determining the inner and outer scaling cutoff, as well as certain scaling parameters. When the phenomena of interest are mono-dimensional, both intense and weak regions scale in the same way, hence there are relatively few parameters involved. Furthermore, the dimension of the space over which the averages are taken is unimportant[1]. In this case, no special care is needed in applying any of the following techniques. However, in the general case where the phenomena are multidimensional, these standard techniques should be supplemented by methods that are specifically designed to study separately weak and intense regions, as well as their dimensional dependence. In Section 10.3.3, we describe a new method that is quite effective for this purpose.

Some general methods for studying scale dependence.

As discussed in the previous section, an important type of scale invariance involves the probability distributions of fluctuations at different scales. In this case, the most obvious way of testing scale invariance is by directly comparing the empirical distributions at different scales. However, clearly,

---

[1] This is true if the dimension is not too small—see Section 10.6

if the probability distributions are scaling, then so are many of their statistical properties such as the variance, high moments, spectra and autocorrelation functions. A number of related techniques exist to test this and the following by no means exhausts the possibilities:

Probability distributions.

The direct analysis of probability distributions is best accomplished by using log-log plots such as those shown in figure 10.2. This method is valuable because not only does it allow us to evaluate H, and the limits to scaling, but it also enables us to estimate α, the hyperbolic intermittency parameter (the negative slope for large ΔX in figure 10.2). Hyperbolic intermittency means that large fluctuations (ΔX) are distributed as follows:

$$\Pr(\Delta X' > \Delta X) \propto \Delta X^{-\alpha} \qquad (10.3)$$

$\Pr(\Delta X' > \Delta X)$ is the probability of a random fluctuation $\Delta X'$ exceeding a fixed ΔX. As discussed in the next section, fluctuations of this type are expected to occur due to the action of cascade processes which concentrate stuff (e.g. energy, matter) into smaller and smaller scales. When the fluctuations are of this type, the phenomena are so intermittent that high-order moments $\langle \Delta X^h \rangle$ diverge (are infinite) for $h \geq \alpha$. For a discussion, as well as much empirical evidence (e.g. wind and temperature data), see Schertzer and Lovejoy (1985a) and Lovejoy and Schertzer (1985a).

Structure functions.

The usual structure functions (discussed, for example, in the turbulence literature), are the "delta" structure functions ($S_d(h, \Delta t)$) defined as follows:

$$S_d(h, \Delta t) = \langle X^h(\Delta t) \rangle \qquad (10.4)$$

where $\Delta X(\Delta t)$ is the fluctuation (or difference, hence "delta") defined earlier. The most commonly used delta structure function is for $h = 2$, the delta variance. Note that in general, scaling implies:

$$S_d(h, \Delta t) = \lambda^{H(h)} S_d(h, \Delta t) \qquad (10.5)$$

When X is mono-dimensional, $H(h)$ is linear in h; however, in the more general multidimensional case (see, for example, Parisi and Frisch, 1985 discussion of "multifractals"), $H(h)$ is more complicated, depending on the various fractal dimensions of the different powers of the X. While this method is useful for studying the difference in scaling of the weak and strong regions (small and large h respectively), in its present form, it is not adapted to studying the effect of varying the dimension over which phenomena are averaged. In Section 10.3, we show how, by studying averages directly, this defect can be remedied.

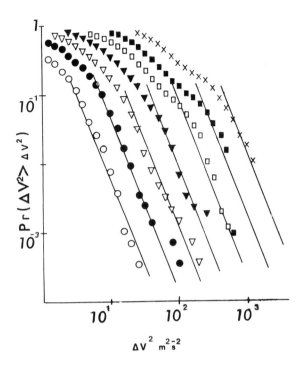

Figure 10.2. An example of a direct analysis of the probability distribution of fluctuations. Δv^2 is the horizontal wind (squared), and fluctuations are examined across layers with Δz (left to right) =50,100,200, 400, 800, 1600, 3200 meters thick. The straight line asymptotes have slopes $(-\alpha_v/2) = 5/2$, and the horizontal spacing is $(H_v)\log 2$ with $H_v \sim 3/5$.

Energy spectra.

The usual energy spectrum of fluctuations $E(k)$ where k is a wavenumber, is also useful. When $E(k) \propto k^{-\beta}$ the spectrum is scaling, and $\beta = 2H(2)+1$, since the delta structure function order two can be written in terms of a definite integral of the spectrum. Hence, for example, the familiar Kolmogorov scaling spectrum $E(k) \propto k^{-5/3}$ law for the spectrum of turbulent wind fluctuations implies $H(2) = 1/3$. Aside from their direct physical interpretation in terms of the energy at different scales, Pentland (1984) has shown that scaling spectra can also be used in image analysis.

R/S analysis.

R/S analysis, or the rescaled range is a robust measure of long-range dependence introduced by Mandelbrot and Wallis (1969). Although it suffers from certain problems (including biases for short series—see Bhattacharya *et al.*, 1983), it can be of use in determining H even when fluctuations about the long-term trend are very large. For geophysical examples, see Mandelbrot and Wallis (1969), and Lovejoy and Schertzer (1985b).

Distribution of areas.

The following techniques are of interest when the data is of 2 or higher dimension. For example, with digital images, areas may be defined objectively by the simply connected regions that exceed a fixed threshold. If the field is scale invariant, then the distribution of such areas a, must be of the form:

$$\Pr(A > a) \propto a^{-B} \qquad (10.6)$$

Where B is a parameter, which is independent of the threshold only if the field is mono-dimensional. The probability distribution must be of this (hyperbolic) form because any other type of distribution would involve a length scale, and hence break the scaling. Geophysical examples include the distribution of ocean islands where $B \sim 0.6$ (Korchak, 1938, Mandelbrot, 1982), and weak rain rates and low clouds, where Lovejoy (1981), Lovejoy and Mandelbrot (1985) find $B \sim 0.75$. Rothrock and Thorndike (1983) find B in the range of 2 for ice flow areas, and Peleg *et al.* (1984) show that hyperbolic area distributions can be used in image analysis and segmentation.

The area-perimeter relation.

The areas defined above will, in general, have very complex perimeters. If we fix the resolution of the data and measure lengths in terms of pixels, and areas by square pixels, then scaling implies:

$$P \propto (A)^{D/2} \qquad (10.7)$$

where D is the fractal dimension of the perimeter (= 1 for smooth shapes such as circles or squares, = 2 for very complicated perimeters that literally fill the plane). Figure 10.3 shows the results of such an analysis on rain and cloud data from radar and satellite images. The value $D \sim 1.35$ seems to hold at least from 160m to 1000km for low, warm clouds, and light rain rates (see Lovejoy, 1982 and Cahalan *et al.*, 1984).

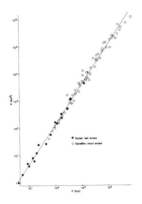

Figure 10.3. The area-perimeter relation for the projections on the Earth's surface of both satellite cloud areas and radar rain areas. In each case, areas and perimeters were obtained by setting a threshold (corresponding to warm, low clouds, and light rain rates respectively). The areas and perimeters correspond to pixel sizes of 1km.

Other methods.

Other methods of data analysis, classification and compaction which exploit the idea of scale invariance include Nguyen and Quinqueton (1982) use of Peano curves, and Bell *et al.* (1985) use of tesselations (see Chapter 8), Dutton's (1981) image enhancement, as well as the straight-line sampling technique described in Lovejoy *et al.*, (1983) and Lovejoy and Schertzer (1985c).

A multidimensional analysis technique: the integral structure function.

A new technique specifically designed for studying the scale dependence of both weak and strong regions (low and high powers of the field), as well as their dependence on the dimension over which they are averaged (D_A), is based on what we call the integral structure function ($S(h, L, D_A)$). Aside from its theoretical interest, it is of direct use in remote sensing applications because it is necessary not only in comparing remotely sensed data at different scales, but also data averaged over different dimensions. It is therefore required in the calibration process.

The general D_A-dimensional function is defined as follows:

$$S(h, L, D_A) = (L^{-D_A})\langle (\int \cdots \iint_{L^{D_A}} (X(\underline{r}))d^{D_A}\underline{r})^h \rangle \qquad (10.8)$$

where L is the size of the D_A dimensional hypercube over which the averages are taken. Scale invariance implies:

$$S(h, \lambda L, D_A) = \lambda^{-p(h,D_A)} S(h, L, D_A) \qquad (10.9)$$

where λ is our usual enlargement ratio, and $p(h, D_A)$ is a function not only of h, but also of D_A. Note that D_A is not restricted to being a line, plane, or volume.[2] Averages can be taken over any fractal set, a point that will be made clear in our discussion of ground station networks in Section 10.5. The minus sign is introduced in the definition because for large h, averages will decrease with increasing scale (averaging over larger scales smooths out the rare, large fluctuations that contribute significantly to the averages). As we increase h, keeping D_A fixed, more and more of the total contribution to the integrals is due to small, but very intense regions. $p(h, D_A)$ therefore gives us direct information about how the intense and weak regions vary as the scale over which they are averaged increases. In Section 10.6, we explain why this function can also be very sensitive to D_A, both by examining a simple multidimensional model, as well as by empirically evaluating p using radar data in the rainfield with D_A = 1, 1.5, 2, 3, 4.

10.4. A Simple Monodimensional model with strong intermittency.

The importance of stochastic models.

We have briefly outlined the empirical evidence (such as the abundance of power law spectra) as well as the theoretical arguments (symmetry principles, GSI), to the effect that the outer scale of the scale invariant regime of many geophysical quantities is of the order of 1000km or more. If this is true, then remotely sensed data are generally averaged over distances which are well within the scaling regime. Averaging is therefore typically performed over a wide range of highly variable structures. Any interpretation of averages therefore involves assumptions about the sub-resolution scales. Unfortunately, phenomena are often implicitly considered to be uniform over these scales. This uniformity assumption is usually justified (e.g. Box and Jenkins, 1970) by supposing that the outer scale (and hence the decorrelation distance) is very small, and that averaging occurs over much larger scales. Limited scale dependence arises because increasing scale increases the number of independent samples, thereby decreasing the sampling error. The true ensemble averages are considered to be independent

---

[2] The extension of the integration to general fractal sets is conceptually simple, but involves mathematical niceties outside our present scope.

of scale. Conversely, within the scaling regime, correlations are strong, and we require the next level of approximation in which the extreme variability is recognised, but is treated only as regards the scale dependence of averages. A still higher level of approximation recognises not only the scale dependence, but also the dimensional dependence of the averages. Both the latter levels of approximation are quite general—actual data analysis requires much more specific assumptions, and this is best accomplished by explicit (stochastic) models of the behaviour in the scaling regimes.

Stochastic models in geophysics have been developed for some time, notably in hydrology (e.g. the review by Waymire and Gupta, 1981). Scaling, fractal models are however more recent (e.g. Mandelbrot, 1975, 1982, Voss, 1983, Andrews, 1980, 1981, Kagan and Knopoff, 1981, Lovejoy and Mandelbrot, 1985, Waymire, 1985 and Fournier et al., 1982). All these models are of the intermediate mono-dimensional type, and are useful in remote sensing because they can provide at least crude models of the sub-resolution scales. These mono-dimensional models are discussed below; they may be sub-divided according to whether their intermittency (variability) is weak (quasi-Gaussian) or strong (fat-tailed or hyperbolic). The quasi-Gaussian case is by far the simplest because all the properties of the field are determined by a single scaling parameter (e.g. the spectral exponent, or the fractal dimension). Models of this type for the Earth's relief have recently been developed (mainly for computer graphics applications), and can be found in Voss, 1983, and Fournier et al., 1982 (see also Goodchild, 1980, 1982). They have already proven useful in the sampling and calibration problems involved in the remote sensing of terrain (Leberl, 1985).[3]

In meteorology, fluctuations are far more intermittent than gaussian models can allow for, and we require not only scaling but also strong (hyperbolic) fluctuations characterised by the intermittency exponent α. The recognition of this lead to the development of hyperbolically intermittent (but still mono-dimensional) models for both the isotropic case (Lovejoy and Mandelbrot, 1985), and for the more general case involving differential stratification and rotation (Lovejoy and Schertzer, 1985b,c). In the appendix, we give a brief description of the simplest of such models which is called the Fractal Sums of Pulses (FSP) process, which is produced by adding together a large number of elementary shapes (the pulses). Figure 10.4 shows an example.

---

[3] In this context, it is probably worth mentioning a recent paper by Mark and Aronson (1984) finding that the fractal dimension, (and hence the scaling) only holds over very short ranges. Actually, these authors only tested how the amplitudes of fluctuations in vertical elevation of the Earth's surface, depend on horizontal scale—they did not test the scaling of horizontal structures which is at issue here. Their observations only test a particular mono-dimensional model called the fractional brownian surface model. We interpret their analysis as indicating that mono-dimensionality (rather than horizontal scaling) should be abandoned: in fact, multidimensional models predict broadly the same sort of behaviour as shown by their data.

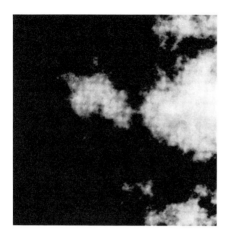 

Figure 10.4. (a) An example of an isotropic FSP process on an 800 × 800 point grid, with whiteness proportional to the log of the rain rate. 640,000 elementary (annular) pulses, with $H = 1/\alpha = 3/5$ were used (the parameters that correspond the closest to the real rain data). (b) An anisotropic FSP model of a vertical rain cross section, showing increasing horizontal stratification with increasing scale on a 400X400 point grid, illustrating generalised scale invariance.

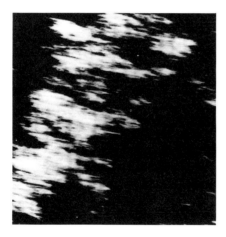

Figure 10.4c. An FSP model of a horizontal rain/cloud field with differential rotation to simulate the Coriolis force (see Lovejoy and Schertzer 1985c for details).

10.5. The Dimension of Surface Network and the need for Remote Sensing.

The dimension of the Canadian surface network is about 1.5.

Strictly speaking, the problem of *in situ* measurements is outside the domain of remote sensing. However, no remote sensing technique can be developed without some form of ground truth. We must therefore consider the fact that, in general, remote and *in situ* sensors measure properties of the fields averaged not only over different scales, but also over different dimensions: for example, satellites will typically sample and average over the Earth's surface ($D = 2$), whereas the ground (calibration) network is primarily concentrated in urban areas, and is thus—as we show below—a highly non-uniform fractal set with $D < 2$.

The reason for the dimensional dependence of averages of multidimensional phenomena is easy to understand. Consider the co-dimension (C) of a set defined by the difference between the Euclidean dimension of the space in which the set is embedded (E) minus the dimension of the set itself (D): $C = E - D$. The general geometric theorem on the intersection of sets with different co-dimensions (C_1, C_2) then states:

$$C_i = C_1 + C_2 \qquad \text{for } C_i < E, \text{ otherwise } C_i = E \qquad (10.10)$$

where C_i is the co-dimension of the intersection set. If the intersection set has dimension zero ($C_i = E$), then it is empty: i.e. with probability one, it contains no points.

This result is immediately applicable to the sampling problem: if an infinite measuring network has co-dimension C_m, then any phenomena with dimension $D_p < C_m$ cannot (with probability unity) be detected. In real measuring networks which have a finite number of stations, and thus an inner and outer scale, the problem is compounded by the inability to detect phenomena much smaller than the minimum.

The dimensional dependence of averaging puts remote sensing in an entirely new perspective. Traditionally, remote sensing has been considered necessary simply in order to increase the sampling rate and density. While this is still true, we now see that remote sensing is necessary for an entirely different reason: it is the only way to increase the dimension of the averages (even to 4, in the case of radar data—see Section 10.6), so as to detect the most intense regions. This is important because although these regions are the most sparse (they have the lowest dimension), they will typically contribute significantly to the average fluxes, energies and other quantities of interest.

In order to test these ideas, we examined the Canadian surface meteorological network (see figure 10.5 for a map showing their geographical locations). In spite of efforts to achieve a uniform distribution of stations over the entire area of Canada (and hence to achieve $D_m = 2$), it is clear that the actual distribution is tightly clustered around the major population centres. Clustering of this sort is typical of a set of points with dimension less than that of E (i.e. with $C > 0$).

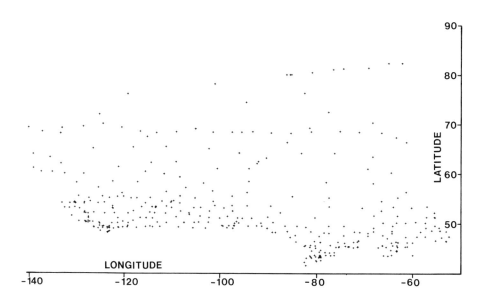

Figure 10.5. The locations in latitude and longtitude of the 414 stations in the Canadian meteorological measuring network showing their high degree of non-uniformity. As expected, most are clustered near the major cities and the US border. The dimension (D_m) is ~ 1.5.

Many empirical methods of estimating the dimension of a set of points can be devised. The method described below has been widely used to estimate the dimension of strange attractors (e.g. Hentschel and Procaccia, 1983) and Nichols and Nichols (1984) for another geophysical application). For each station, determine the number ($n(L)$) of other stations within various radii L of the point. We can thus determine the average over all the stations, $\langle n(L) \rangle$. If the stations are distributed uniformly over an area ($D_m = 2$), then $\langle n(L) \rangle \propto L^2$: in general, if they are distributed over a set dimension D_m, then a definition of the distance between stations such that increasing distances by factor of λ will increase the number of points in the uniform case by λ^2. A metric which has this property is clearly the square root of the area of the spherical cap defined by the two points. Using this definition, we are lead to the following formula:

$$L = r\sqrt{8(1 - \cos\theta/2)} \qquad (10.11)$$

where r is the radius of the Earth, and θ is the angle subtended by the two points at the Earth's centre. Note that we have dropped the constant factor $\sqrt{\pi/4}$ so that for small θ, the formula reduces to the usual great circle distance ($= r\theta$).

Figure 10.6 shows $\langle n(L) \rangle$ calculated using the above definition of L, for the 414 station network in Canada. Each station defines 413 distances, therefore, there are $414 \times 413/2 = 85491$ independent values that go into the histogram for calculating $\langle n(L) \rangle$. From the figure, we can see that over nearly two orders of magnitude in L (roughly between 30 and 1000km), $\langle n(L) \rangle \propto L^{1.5}$, showing that over this range, $D_m \sim 1.5$. As expected, at scale sizes of the order of the size of Canada, the $L^{1.5}$ behaviour breaks down, data from a larger region is necessary to determine the true outer scale. As to the behaviour at small scales, it indicates an overabundance of extremely close stations. Actually, figure 10.6 underestimates the true number of very close stations since the geographical locations were only specified with an accuracy of 1' of arc ($\sim$ 2km). Whenever several stations had nominally the same location, only one of them was used. This is necessary in order to avoid meaningless zero distance cases. This procedure resulted in the exclusion of 23 stations from an original sample of 437.

Simulations of measuring networks with $D_m < 2$.

At sufficiently small and large distances, the relationship $\langle n(L) \rangle \propto L^{D_m}$ must breakdown—if only because there is only a finite number of points in the sample. It is therefore of some interest to compare the empirical $\langle n(L) \rangle$ function with that obtained from limited size simulations with the same dimension.

There are many ways to generate sets of points with arbitrary dimension, some of which are discussed in Mandelbrot (1982), (see also Lovejoy and Schertzer, 1985a for another geophysical example). When $D < 2$ (as it is here), the simplest is the "Levy flight", which is produced as follows. Starting at a random point, choose a vector with random orientation and length (L), distributed as $\Pr(L' > L) \propto L^{-D}$ (which represents the probability of a random length L' exceeding a fixed L). If the number of points tends to infinity, and L tends to zero, then the resulting set has dimension D.

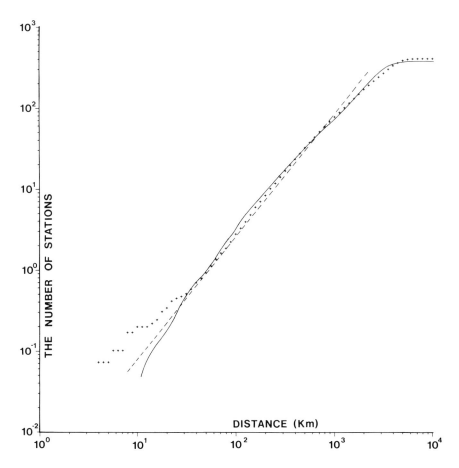

Figure 10.6. The average number of stations ($\langle n(L) \rangle$) within a radius L of a given station for the stations shown in figure 10.5. The straight (dashed) line is the function $\langle n(L) \rangle \propto L^{D_m}$ which would be obtained from an infinite network with $D_m = 1.5$. The solid line shows the effect of finite sample size as determined by simulations of a fractal network with $Dm=1.5$, with only 414 points. The plus signs represent the empirical function obtained from the Canadian meteorological network, showing good agreement with the simulation down to $\sim 30km$.

When this method is used to generate a set of 414 points, with $D = 1.5$, the $\langle n(L) \rangle$ function is very near the $L1.5$ line. Both the small deviations from this line due to the finite sample size, as well as the flattening off at the outer scale ($\sim$ 1000km), are well reproduced (figure 10.6). It is only for $L < 30$km that a substantial difference between the simulation and the actual data occurs.

To fully appreciate the significance of our results, it should be compared to the standard method for analysing ground networks. The usual approach assumes the stations to be "nearly" uniform with $D_m = 2$ and obtains the inner scale (resolution) of the network by taking the square root of the average area per station. In Canada, this calculation yields a resolution of approximately $\sqrt{1 \times 10^7/400} \sim 160$km which is a factor ~ 5 larger than the true resolution. Furthermore, the effective dimension of the network is considerably less than 2, implying that some phenomena (with $D_p < 0.5$), have a probability zero of being detected. To detect such phenomena (which include the most intense parts of atmospheric fields), it is necessary to develope remote sensing techniques with $D_m = 2$ (or, higher). In the next section, we show that even when the dimension of the phenomenon is large enough to enable it to be detected, that all its basic statistical properties still depend on D_m.

10.6. Cascade Processes and Multidimensionality.

Cascade processes.

Ever since Richardson (1922), the idea of cascade processes transferring energy from large to small scales has been central to theories of fully developed turbulence. By assuming a homogeneous (translationally invariant), self-similar cascade (such as that sketched in figures 10.1a,b), Kolmogorov, 1941 obtained his famous $E(k) \propto k^{-5/3}$ energy spectrum for the energy ($F(k)$) of fluctuations in the wind at wavenumber k. Soon afterwards, Batchelor and Townshend (1949) underlined the importance of the "spottiness" or intermittency of turbulence. One early attempt to account for intermittency was Kolmogorov's (1962) proposal that the energy flux was log-normally distributed[4].

Attempts to model theoretically intermittency lead, via the work of Novikov and Stewart (1964), and Yaglom (1966) to Mandelbrot's (1974) general cascade scheme. Mandelbrot argued that the spottiness reflected the fact that the active regions (the support of turbulence) were concentrated on a compact set with $D < 3$. In addition to a fractal support, this scheme featured hyperbolic distribution of the energy flux. Unfortunately, in later work such as Frisch et al. (1978), only the simplest mono-dimensional (" β-model") was considered. This case is very special—for example, it does

---

[4] Log-normal distributions have much more probably concentrated in their tails than the more usual statistical distributions which are asymptotically exponential. We therefore consider them long tailed. However, all their statistical moments exist (although barely), hence we still distinguish them from from truly fat-tailed hyperbolic distributions.

not involve the divergence of the high-order statistical moments.

Schertzer and Lovejoy (1983, 1985a) pointed out that in general the model involves not one, but a whole hierarchy of dimensions of support; the dimension being lower for the more intense regions. The related phenomenon of the divergence of moments was shown to occur whenever the dimension over which the process was averaged was sufficiently small. Schertzer and Lovejoy (1985a) proposed a generalisation to anisotropic cascades, necessary for example to account of the stratification introduced by gravity, and the rotation associated with the Coriolis force. They also proposed (with empirical support from the wind, temperature, rain and other fields), that hyperbolic intermittency is a general feature of atmospheric fluctuations. More recently Schertzer and Lovejoy (1985b) outlined a formalism (called generalised scale invariance, GSI), which makes possible analysis and simulations of multidimensional measures without the artificial grid involved in the earlier models. Hentschel and Procaccia (1983) and Grassberger (1983) independently showed that strange attractors involve a hierarchy of dimensions. Related work on multiple dimensions are Parisi and Frisch (1985) and Mandelbrot (1984a). It now seems that multidimensionality is quite natural: *a priori*, it would seem to be much more so than the homogeneous mono-dimensional case—hence, even the cartographic data upon which models of random topography are based should be re-analysed with a view to characterising the whole hierarchy of dimensions probably involved[5] (the same is true of Lovejoy's (1982) analysis of cloud and rain perimeters). In any case the phenomenon of multidimensionality and anisotropic scale invariance (GSI), clearly establishes that scale invariance is a more basic concept than fractal dimensionality: not only can scale invariance lead to a whole hierarchy of dimensions, but the latter can even be position-dependent (as it is in non-linear GSI—Schertzer and Lovejoy, 1985b). This could be important for models of the Earth's topography since, for example, it enables us to accomodate—without invoking a breakdown of the scaling symmetry—the evidence (c.f. Goodchild, 1982) that different parts of the topography are characterised by different fractal dimensions[6]. In the next section, we illustrate multidimensionality by considering two special cases.

---

[5] This is especially true of Mark and Aronson's (1984) empirical results.

[6] This is quite different from Takayasu's (1982) idea of *scale* dependent fractal dimensions in which the scaling symmetry is broken (albeit in a continuous manner).

A simple multidimensional model.

Consider the cascade process outlined in figures 10.1a,b, whereby an eddy at scale L_n (figure 10.1a) divides by a factor λ (here, $\lambda = 2$) into sub-eddies, size $L_n = L_{n-1}/\lambda$. At each stage, the energy (ϵ_{n-1}) is redistributed by random factors W into the sub-eddies as follows:

$$\epsilon_n \langle i \rangle = W_i \epsilon_{n-1} \quad (10.12)$$

where i indexes the λ^D sub-eddies (D is the dimension of space, in figure 10.1, $D = 2$). The condition on the independent and identically distributed W that ensures conservation of energy from one scale to another (i.e. that the energy is a scale invariant measure), is $\langle W \rangle = 1$. When this process is iterated $n \to \infty$, the result is the (mathematically) singular measure originally proposed by Mandelbrot (1974), and discussed in Kahane (1974), and Peyrière (1974). This means that the physically meaningful quantities are the spatial averages (over scale L_n denoted ϵ_n) over dimension DA. Figure 10.7 shows a sample of this process in two dimensions.

Various properties of these measures are discussed in the above references. In particular, in Schertzer and Lovejoy (1985a), two simple cases were used to illustrate the basic features of the process. Both can be described by a two-state Bernoulli process for W. The first, the "β-model" is obtained by:

$$Pr(W = 0) = 1 - \lambda^{-C_s} \quad (10.13)$$
$$Pr(W = \lambda^{C_s}) = \lambda^{-C_s} \quad (10.14)$$

which yields either "dead" or "active" regions, the latter having co-dimension C_s. At each stage in the cascade, the energy of all the active regions is the same, increasing as $n \to \infty$, the energy of the inactive regions zero. The result is the trivial fractal measure in which the active regions are a fractal set, with each point carrying an infinite energy flux. The second case studied in Schertzer and Lovejoy (1985a), the "α-model" is obtained by choosing:

$$Pr(W = \lambda^{C_\infty}) = \lambda^{-C_\infty h_\infty} \quad (10.15)$$
$$Pr(W = a_1) = 1 - \lambda^{-C_\infty h_\infty} \quad (10.16)$$

where $a_1 = (1 - \lambda^{-C_\infty(h_\infty - 1)})/(1 - \lambda^{-C_\infty})$ with $h_\infty > 1$ and $C_\infty > 0$, are parameters of the process. C_∞ is the co-dimension of the most active regions, with the weaker regions having $C_s < C_\infty$. The most active regions are best studied by considering the measures defined by ϵ^h. Schertzer and Lovejoy, 1985a obtain an explicit formula for the dimension $D_s(h)$ of the support of these measures. Furthermore, when the measure ϵ is averaged over a set dimension D_A which is sufficiently small, its high-order moments diverge.

Figure 10.7. An example of the α model of turbulent cascades with two sub-eddies per eddy, with 9 cascade steps on a 512 × 512 point grid with $D_\infty = 1$, $h_\infty = 2$. The logarithm of the energy flux ϵ is indicated on a grey scale with weak regions white (they are not however completely dead) and intense regions black. If we fix a threshold, the set of points exceeding it will look something like that shown in figure 10.1b (i.e. $D_s < 2$), with the difference that D_s is now a decreasing function of the threshold. The model is therefore multidimensional.

A numerical study of spatial averages of multidimensional measures: the behaviour of the structure function exponents $p(h, D_A)$.

Analysis of the fractal measure ϵ described above is a two-stage process. In the construction stage, eddies are successively broken up into sub-eddies, and the energy redistributed down to infinitely small sizes. Without going any further, many properties of the measure such as the divergence of high moments and its multidimensional nature, may be deduced. However, full analysis, particularly of the spatial averages, requires that the cascade be retraced in the inverse direction, from small to large scales. Specifically, we wish to study the behaviour of spatial averages as we go from sub-eddies to eddies. The simplest way to do this is via numerical determination of the integral structure function exponent $p(h, D_A)$ (c.f. Section 10.3.3), determined from the equation $p(h, D_A) = \log_\lambda(S(h, 1, D_A)/S(h, \lambda, D_A))$. In this subsection, we analyse the properties of $p(h, D_A)$ by considering both the mono-dimensional β-model, and a more interesting model called the parabolic model[7].

We first note, that by construction, $\bar{\epsilon}$ is scale invariant, hence $p(1, D_A) = 0$ for all D_A. Furthermore, $p(h, D_A)$ is zero for $h = 0$, is negative for $0 < h < 1$ and is positive for $h > 1$. Now, consider the shape of $p(h, D_A)$ for the β model where ϵ is non-zero only over a fractal set with co-dimension C_s. If $D_A < C_s$, then, according to the theorem on intersections (Section 10.5.1), the intersection of the averaging region and the fractal has dimension 0, (i.e. the set is empty) and all the spatial averages depend only on D_A (not C_s). However, if $D_A > C_s$, then the co-dimension of the intersection set with respect to D_A is constant $(= C_s)$[8]. In this case, the properties of the spatial averages depend only on the fraction of the averaging space (at scale λ) that are active: $\lambda^{D_i}/\lambda^{D_A} = \lambda^{-C_s}$. Hence, if $D_A > C_s$, $p(h, D_A)$ is independent of D_A. The exact result (for $h > 0$) is:

$$p(h, D_A) = C_s(1 - h) \quad \text{for } D_A > C_s \quad (10.17)$$
$$p(h, D_A) = D_A(1 - h) \quad \text{for } D_A < C_s \quad (10.18)$$

We therefore conclude that a basic characteristic of mono-dimensional fields is that $p(h, D_A)$ is linear in h and for $D_A > C_s$ is independent of D_A. Non-linear $p(h, D_A)$ with respect to h, and strong dependence on D_A are clear indicators that the variablity is of the multidimensional type.

Now consider the parabolic model where W is log-normally distributed. Figure 10.8 shows the functions $p(h, 1)$ and $p(h, 0.8)$[9] for a parabolic process modelled in one dimension. These functions have been estimated from

---

[7] Mandelbrot (1974) shows that the basic function determining the characteristics of ϵ is $\log\langle W^h\rangle$. In the "β" and "α" models described above, these functions are respectively linear and asymptotically linear in h. A somewhat more interesting model to compare with rain data is the case where this function is parabolic in form, as it is when W has a log-normal distribution (see Mandelbrot, 1972). We call this case parabolic rather than log-normal because we wish to avoid confusion with log-normal models of $\bar{\epsilon}$ (e.g. Kolmolgorov, 1962). In the parabolic model, it is not $\bar{\epsilon}$, but rather W that is log-normally distributed ($\bar{\epsilon}$ is in fact hyperbolically distributed).

[8] This is easy to show. Using the notation of Section 10.5.1:

$$C_i = C_A + C_S \quad D_A > C_S$$
$$\Rightarrow E - D_i = E - D_A + C_S$$
$$\Rightarrow D_A - D_i = C_S$$

[9] The curve for $D_A = 0.8$ was obtained by averaging over a Cantor set with dimension $\log 3/\log 4 \sim 0.8$.

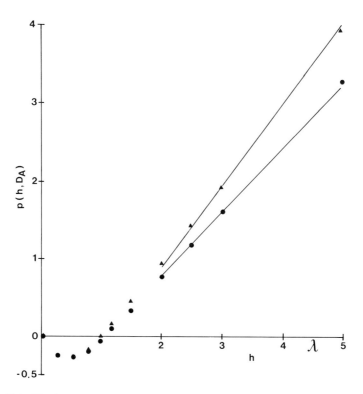

Figure 10.8. The function $p(h, D_A)$ for $D_A = 1$ (triangles), 0.8 (circles), obtained numerically from 64 samples of the parabolic model with two sub-eddies per eddy, and 12 cascade steps (hence length $2^{12} = 4096$ pixels). The straight-lines have slopes $= D_A$ (i.e. $=1$, 0.8 respectively), showing that the most intense regions have dimension zero (see text).

64 samples of 4096 points, (2 sub-eddies per eddy, with 12 cascade steps)[10]. In figure 10.8, note in particular that the large h region has an asymptotic slope of D_A. Simulations show that the asymptotic slope is equal to the co-dimension $C_s(\infty)$ (with respect to D_A i.e. $C_s(\infty) = D_A - D_s(\infty)$) of the most active regions present. In the parabolic model, these regions always have $D_s(\infty) = 0$, hence $C_s(\infty) = D_A$. However, in the α-model,

---

[10] S and hence p are defined as statistical (ensemble) averages, length L. However, in order to increase the rate of convergence, the ensemble averages were estimated by using the rate of convergence, the ensemble averages were estimated by using all the available disjoint samples of length L, (i.e. here, this yields $64 \times 4096/L$ rather than 64). We numerically verified that this procedure does not introduce a bias in $p(h, D_A)$, and furthermore, has the beneficial effect of increasing the rate of convergence.

thresholds can be chosen defining fractal sets with any dimension between 0 and C_∞, hence the slope is C_∞ or D_A, whichever is smaller.

We can now qualitatively account for the behaviour of multidimensional measures. Although the most intense regions are very rare (sparse, low $D_s(h)$), they are sufficiently intense so as to give a large contribution to the averages ($\langle \bar{\epsilon}^h \rangle$); however, the averages with small D_A will exclude all the intense regions lying on fractal sets with $D_s(h) < C_A$, hence, information on the field is lost whenever D_A is less than its maximum possible value (= E, the euclidean dimension of the space in which the process occurs). This shows the importance of averaging over as large a dimension as possible.

Note that as expected, the simulation used in figure 10.8 involves the divergence of moments. In this case, it was found numerically that $\alpha \sim 0.66$.

Multidimensionality in the rain field: the empirical $p(h, D_A)$ is non-linear and depends on D_A.

Of all the geophysical fields, none are known with as high a resolution in the four dimensions of space and time as the radar-determined rain field. For example, the data used in the study described below were from 5 series of Constant Altitude **Z LO**g **R**ange maps (CAZLORs) taken at 3,4,5 km altitudes every 30 minutes from the McGill weather radar observatory, using $70 \times 3 = 210$ CAZLORs in all. The resolution of the data was 375 in azimuth (= 0.96), and 200 downrange elements (from 20 to 200km) logarithmically spaced in order to obtain nearly cubical resolution elements. The entire data set therefore contained $210 \times 375 \times 200 = 1.575 \times 10^7$ elements. Note that the extreme variability of both the cascade models and the rain data means that extremely large samples such as these are necessary. For each element, the radar reflectivity $\bar{Z}$ was recorded at resolution of 4 bits (16 levels) of logarithmically spaced intensities. The radar measures the total backscatter from all the drops within a scattering volume, with an amplitude proportional to the drop volume squared, and with a random phase due to the random positions of the drops. The total integral $\bar{Z}$ is indirectly related to the rain rate ($\bar{R}$), by an approximate formula due to Marshall and Palmer (1948)[11]: $\bar{R} \propto \bar{Z}^{0.6}$. However, for the purposes of this paper, we directly consider Z which avoids the (non-trivial) problem of radar calibration with raingages. In fact, our analysis will show why the latter is so difficult.

First, we confirm (figure 10.9), that $\bar{Z}$ is hyperbolic. For $D_A = 1$, $\Pr(\bar{Z}' > \bar{Z}) \propto \bar{Z}^{-\alpha_Z}$ with $\alpha_Z \sim 1.0$. Thus, $\langle \bar{Z} \rangle >$ is either infinite or barely finite, but the moments $\langle \bar{Z}^h \rangle$ with $h > 1.0$ diverge. It is interesting to note that this is consistent with data from drop volume distributions which

---

[11] The original Marshall-Palmer formula is $Z = 200 R^{1.6}$ with Z measured in $m^{-3}.mm^6$, R in mm/hr.

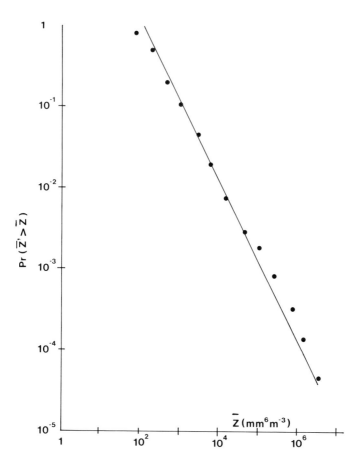

Figure 10.9. The probability $\Pr(\overline{Z}' > \overline{Z})$ *that a random radar reflectivity average* $(\overline{Z}')$ *exceeds a fixed threshold* $\overline{Z}$ *for the sample of radar data discussed in the text. The straight line corresponds to the function* $\Pr(\overline{Z}' > \overline{Z}) \propto \overline{Z}^{-\alpha_z}$ *with* $\alpha_z = 1.0$.

show $Pr(V' > V) \propto V^{-\alpha_V}$, $\alpha_V \sim 2$ (Lovejoy and Schertzer unpublished analyses). These distributions are consistent with each other since $Z \propto V^2$ hence $\alpha_V \sim 2\alpha_2$.

Next, we evaluated $S(h, L, D_A)$ for various values of h, L, D_A (e.g. figures 10.10 a,b). The straightness of these $\log(S(h, L, D_A)/S(h, \lambda L, D_A))$ vs. $\log \lambda$ graphs clearly shows that even at the maximum distance considered (~ 200km), that there is no evidence of a length scale. All moments (except the scale invariant case $h = 1$), are strong functions of λ, clearly showing the

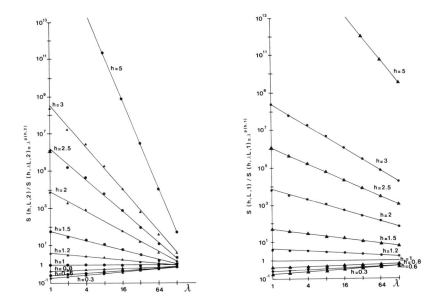

Figure 10.10. (a) The ratio of integral structure functions: $S(h, L, 1)/S(h, \lambda L, 1) = \lambda^{p(h,1)}$ for various powers (h) of radar reflectivity and ratios λ increasing by factors of 2 from 1 to 128, calculated from the data discussed in the text. The averaging was over all disjoint $1-D$ downrange segments of length resolution elements (hence $L = 1$ pixel, 1km on average). The raw reflectivies were normalised by dividing by the mean $=110$ $mm^6 m^{-3}$. The straightness of the lines shows that the rain field has no characteristic scale, and the flatness of the $h = 1$ line shows that Z is scale invariant. The slopes as a function of h yield $p(h, 1)$. Their nonlinear variation is a consequence of the multidimensional nature of the rain field. The point of convergence of the lines is determined by the sample size—see the text. (b) The same as (a), but for 2-D averages (i.e. over all the disjoint X (downrange X crossrange) samples. The difference between 10.10a, 10.10b is a consequence of multidimensionality.

inadequacy of the 0th order approximation of statistical uniformity (independence of scale). In practice, in determining $S(h, L, D_A)/S(h, \lambda L, D_A)$ for ratio λ, L was taken as a single resolution element and the spatial averages of Z were taken over all disjoint coverings of regions with λ^{D_A} elements. For example, with $D_A = 1$, we averaged $\overline{Z}^h$ over consecutive azimuthal sections of length λ elements and took the expectations of the resulting Zh over all the possible disjoint coverings. Note that although elements near and far from the radar have different sizes (because of beam spreading), the method does **not** yield a biased estimate of $p(h, D_A)$ since we compare data at different scales with the same mix of near and far sampling volumes. Any range dependence will therefore effect $S(h, L, D_A)$

and $S(h, \lambda L, D_A)$ in the same way, and thus not alter their ratio $\lambda^{p(h, D_A)}$. A striking aspect of figure 10.10 is that the lines for various h, converge at a point on the λ axis. Numerical simulations show that this point is not physically significant: as the sample size increases, the lines spread apart (without changing their slope, $p(h, D_A)$), and intersect the axis at increasing distances from the origin. This effect is directly related to the divergence of the high moments: in both our model and our empirical calculations we have replaced the true ensemble averages by empirical sums. When moments diverge, the law of large numbers breaks down, and the sums diverge as sample size increases (see the discussion in Schertzer and Lovejoy, 1983a).

Figure 10.11 shows the functions $p(h, D_A)$ for D_A = 1, 1.5, 2, 3, 4 corresponding to a) azimuthal averaging, b) averaging over 1.5 dimensional fractals (to simulate a rainguage network—we used the simulation technique outlined in the previous section), c) azimuth-range averaging, d) azimuth-range-elevation averaging, e) azimuth-range-elevation-time averaging. For DA = 1, 1.5, 2, the range of scales used was 27 = 128 whereas in the DA = 3, 4 cases, the range was limited to a factor of only 3 due to the limited vertical sampling of the radar. This figure shows the inadequacy of the next level of approximation: mono-dimensionality. If the rain field was mono-dimensional, these curves would all reduce to a single straight line (i.e. linear in h, independent of D_A). Clearly, a good model of the rain field (and undoubtedly of many other geophysical fields) must be multi-dimensional. Since little is yet known either theoretically or empirically about multidimensionality, both new methods of data analysis and new, more satisfactory stochastic models are required. Note that exactly the same data was used for calculating all these curves at different resolutions and dimensions, hence the scale and dimension dependence is unambiguous—the results are not artefacts due to the comparison of different samples.

Qualitatively, the figure resembles the model results shown in the previous sub-section, with Z rather than ϵ being an invariant quantity. For large h, all the curves approach asymptotic slopes D_A indicating that the most intense regions of rain, even in 4 dimensions, have dimension 0. Even for the more interesting low h end, the curves are significantly different, underlining the large effect of changing D_A. Of particular interest is the value of $p(0.6, D_A)$ since this is the exponent of most direct relevance to the problem of rainguage calibration of the rainfield. These results show that averaging the rainfield (i.e. $Z^{0.6}$) over larger and larger scales yields a larger and larger mean ($p(0.6, D_A)$ is negative)—Z is apparently the true scale invariant quantity, not R. Incidentally, scale dependent effects have puzzled radar meteorologists for some time (see e.g. Rogers, 1971, or Drufuca and Zawadzki, 1983). This clearly shows the difficulty of calibrating a radar with a rainguage network—the two have different values of D_A, L, as well as h.

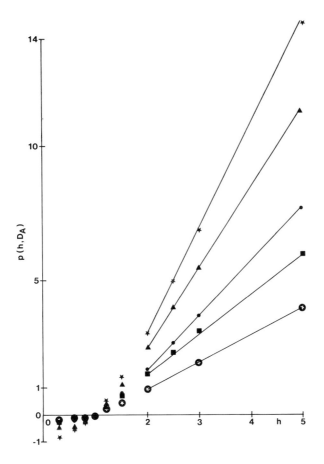

Figure 10.11. The structure function exponents $p(h, D_A)$ *for (symbols bottom to top respectively),* $D_A = 1$ *(from figure 10.10a),* $D_A = 1.5$ *(using simulated fractal rainguage networks),* $D_A = 2$ *(from figure 10.10b),* $D_A = 3$ *(space),* $D_A = 4$ *(space-time). The straight lines for large h have slopes* D_A *which indicate (see simulation, figure 10.8) that the most intense regions have dimension zero.*

10.7. Conclusions.

Many geophysical quantities are extremely variable over a wide range of space and (or) time scales. Over these ranges (which at least in some cases, span $\sim$ 1mm to $\sim$ 1000km in scale), we argue that the statistical properties of the fluctuations respect a symmetry principle called generalised scale invariance in which the small and large scales are related by a magnification, coupled with a geometrical transformation (such as a differential rotation or stratification). Theoretically, GSI is the simplest hypothesis linking structures at different scales, since it doesn't involve symmetry breaking mechanisms (i.e. characteristic lengths or times). Empirically, scale invariance can be inferred by a variety of means (see Section 10.3), including the familiar use of energy spectra.

Over ranges of scale within the scaling regime, the usual statistical methods based on exponential decay of correlation functions are inappropriate. The phenomena are fundamentally non-uniform (intermittent). This leads to the first basic point of this chapter—that in general, all the statistical properties of the spatial and temporal averages depend in a simple, power law manner on the scales over which they are averaged. Sufficiently small moments increase with scale, whereas, the high moments decrease. The intermediate power where the moments are independent of scale defines scale invariant quantities that are physically significant because they are conserved from one scale to another. In Section 10.6, we show this empirically for the rain field.

The scale dependence of averages requires that the active regions (defined by a threshold) of a quantity are distributed over a fractal set whose dimension is less that that of the space available. The simplest models of such behaviour are mono-dimensional, the intense and weak regions are characterised by the same fractal dimension. Mandelbrot (1975, 1982) and Voss's (1983) fractional brownian surface models of the Earth's relief are probably the most familiar of this type. In Section 10.4, we outline a different mono-dimensional model, called the FSP process (Lovejoy and Mandelbrot, 1985), that has much stronger variability and is useful in modelling rain and cloud fields. The discussion of these models brings out the second major point: the necessity to develop stochastic models of the sub-resolution scales of remote sensing devices. An understanding of extreme variability requires the development of stochastic fractal models, on the one hand to make the theory more precise, and on the other hand, in order to understand the details of the sampling, averaging and calibration of remotely sensed quantities.

The third and last basic point in this chapter, is that remotely sensed averages are not only scale dependent, but are also dimension dependent. Even if we fix the scale, statistical properties will still be different if we compare e.g. averages along a line ($D_A = 1$), or along a plane ($D_A = 2$). The possibility of dimensional dependence is due to the recent discovery of (mathematical) measures with multiple fractal dimensions (abbreviated to "multidimensional"). Multidimensional phenomena are scale invariant, but with the more intense regions characterised by lower fractal dimensions than the weaker regions. Averaging over dimensions less than that of the full space/time in which the phenomena are embedded implies that in general, some of the most intense regions (with the lowest dimension), will not be included in the averages. As we average over sets with lower and lower dimension, more and more of the intense regions are excluded, thereby modifying all the statistical properties.

In Section 10.6, we illustrate both the spatial and dimensional dependence of the rainfield (in 1, 1.5, 2, 3, 4 dimensions) by using a new analysis technique we call the integral structure function. Even aside from its theoretical interest, knowledge of this function is necessary whenever we compare data averaged over different scales and dimensions, (e.g. during calibration).

The significance of the dimensional dependence of averages is underlined by our study (Section 10.5) of the dimension of the Canadian surface meteorological network. Over the range of ~ 30 to ~ 1000km (the scale size of Canada), the network (regarded as a set of points) has dimension ~ 1.5 (rather than the value 2 which it would have if it were uniform), indicating a clustering of stations (at all scales) about the various population centres. This observation puts remote sensing in a new perspective: not only is it necessary in order to increase the density and rate of sampling of geophysical quantities: it is also necessary simply in order to increase their **dimension**. For example, if the ground station network is regarded as an infinite set of points with $D_m = 1.5$, distributed over a plane, then any phenomenon (even if infinitely large) with $D_p < 0.5$ cannot (probability=1) be detected.

Many of the analytical tools, models, and theoretical developments outlined in this Chapter, grew out of recent work in chaos, fractals and especially hydrodynamic turbulence. What these fields have lacked in the past, are copious sources of relevant data spanning many orders of magnitude in scale. On the other hand, the remote sensing community has lacked systematic methods for dealing with extreme variability. We believe that it is now possible to end this dichotomy: for the first time, we have not only the quantity and quality of data and the necessary computing capacity, but also a growing variety of methods for systematically analysing and modelling. The consequences will undoubtedly be important not only for remote sensing, but also for turbulent and chaotic systems.

10.8. Appendix: The Fractal Sums of Pulses Process.

Consider a function $R(t)$ formed as the sum of rectangular pulses of random heights representing a rainfall intensity increment ΔR, random widths representing rainfall duration ρ and random centres distributed along the t axis (by a Poisson process rate ν). If $\Pr(\rho' > \rho) \propto \rho^{-1}$ and $\Delta R = \pm \rho^{1/\alpha}$, then the resulting process is scaling, parameter $H = 1/\alpha$, because increasing length scales by the factor λ only changes the intensity by the factor $\lambda^{1/\alpha}$. Note that this implies $\langle \rho \rangle \to \infty$ which is quite different from the usual strongly scale dependent processes obtained when $\Pr(\rho' > \rho)$ is an exponential or other scale dependent function (see for example Waymire and Gupta, 1981). The scaling may be understood by the fact that the number of pulses over an interval of length ℓ, whose length exceeds ρ, is $\ell \Pr(\rho' > \rho) = \ell \rho^{-1}$ which is invariant under the scale transformation $\ell \to \lambda \ell$, $\rho \to \lambda \rho$. By construction, the increments of the process are hyperbolic with exponent α.

To produce rainfield simulations in 2 dimensions (x, y), the rectangular pulses are first replaced by upright cylinders with circular bases (area A), with height (ΔR) related to A by $R = \pm A^{1/\alpha}$ and with A chosen so that $\Pr(A > a) \propto a^{-1}$. Second, the circular bases of the cylinders are replaced by equal area annuli such that the unit annulus (area $\sqrt{\pi}$) has an outer radius Λ (hence inner radius = $\Lambda' = \sqrt{\Lambda^2 - 1}$, and third, the sharp pulse edges are replaced by a smoother function. The final pulse shape used, (see the examples in figures 10.4a,b,c) for pulse size ρ, (ρ = radius here) was:

$$\Delta R\, e^{(-((u^2/\rho'^2)-\delta^2)/\sigma^2)^{2S}}$$

where u is the distance from the pulse centre, $R = \pm a^{1/\alpha}$ the amplitude, and δ is the radius of the annulus centre (= $1/2(\Delta^+ + \Delta)$) and $\sigma = 1/2(\Delta^\times - \Delta)$ is the annulus width. S is a parameter that was introduced to vary the pulse smoothness: $S \to \infty$ yields sharp-edged rectangular pulses. In practice, we generally took $S = 2$. Furthermore, only pulses between some outer and inner cutoff (ρ_0, ρ_i respectively) were used. In the simulations shown here, ρ_0 was taken to be equal to 3 times the size of the simulation "window" and ρ_i was taken to be 1 pixel. We also take $\nu = 1$. Δ was introduced because it allows us to vary the texture of the resulting field—the value $\Delta = 1.2$ was found to be the most visually realistic (see Lovejoy and Mandelbrot, 1985 for illustrations with varying Δ and Mandelbrot, 1984b for mathematical details). Note that if the fluctuations were of the exponential type, then the basic pulse shape would not matter: as the number of pulses increased, the field would eventually tend to a Gaussian limit depending only on H. Here, on the contrary, the fluctuations are so strong that the largest pulse can dominate the others, hence its shape subtly influences the characteristics of the process (this phenomenon was called the "Noah effect" by Mandelbrot and Wallis (1968) after the extreme fluctuation responsible for the Biblical Flood). This mechanism allows the FSP process to produce a very rich phenomenology of shapes in accord with the diversity of real atmospheric fields. If this process is generalised to 3 (or higher) dimensions (e.g. (x, y, t)) by, for example, replacing the annuli with spherical shells), we obtain a model of the temporal evolution of the horizontal field, which is realistic over ranges in which the spatial and temporal statistics are similar (the Taylor hypothesis of "frozen" turbulence). In the atmosphere, this approximation may not be too bad for distances up to a thousand kilometers, and time periods of a day or so (see the discussion in Lovejoy and

Schertzer, 1985a).

No matter what the dimension, the function R cannot be immediately interpreted as a rain rate because the value of R is almost surely negative at some points. In practice, a threshold R_T is set, and the rain rate is measured as the difference $R - R_T$, and the remaining negative values are reset to zero (shown as black background in figure 10.4). The logarithm of the result is then shown on a grey scale with white being the most intense. The resemblence of these FSP simulations and real clouds is striking, especially considering that no effort has been made to take into account the all important multiple scattering processes. Apparently, at least to first order, the light scattered by clouds depends on the logarithm of the total water substance (the optical depth), which is closely related to the rain rate integrated along a given direction.

Acknowledgements.

We greatly benefited from both discussions and encouragement from Peter Muller and Geoff Austin. We also acknowledge discussions with R. Cahalan, D.K. Lilly, B. Mandelbrot, R. Peschanski, J. Peyrière, M. Rafy.

References.

Andrews, D.J. 1980, A stochastic fault model 1. Static case. *Journal of Geophysical Research*, 85 B7, 3867–3877.

Andrews, D.J. 1981: A stochastic fault model 2. Time-dependent case. *Journal of Geophysical Research*, 86 B11, 10821–10834.

Batchelor, G.I., A. A. Townsend, 1949, The nature of turbulent motion at large wave numbers. Proc. Roy. Soc., A199, 238–250.

Bell, S., B., Diaz, F. Holroyd, 1985, Tesseral addressing for rasters, vectors and point data. (in this book).

Bhattacharya, R.N., V.K. Gupta, E. Waymire, 1983, The Hurst effect under trends. *J. Appl. Prob.*, 20, 649–662.

Bills, B.G., M. Kobrick, 1985, Venus topography: a harmonic analysis. *Journal of Geophysical Research*, 90 B1, 827–836.

Box, G.E.P., G.M. Jenkins, 1970, *Time series analysis: forecasting and control.*, Holden-Day, San Fran., 553pp.

Bradbury, R.H., R.E. Reichelt 1983, Fractal dimension of a coral reef at ecological scales. *Mar. Ecol.-Prog. Ser.*, 10, 169–171.

Burroughs, P. A. 1981, Fractal dimensions of landscapes and other environmental data. *Nature*, 294, 240–242.

Cahalan, R.,Wiscombe, Joseph, 1984, Fractal clouds from LANDSAT imagery. Preprints, NASA, Greenbelt, Md.

Deschamps, P.Y., R. Frouin, L. Wald, 1981, Satellite determination of the mesoscale variability of the sea-surface temperature. *J. Phys. Ocean.*, 11, 864–870.

Dutton, G.H., 1981, Fractal enhancement of cartographic line detail. *Amer. Cart.*, 8, 23–40.

Drufuca, G., I.I. Zawadzki, 1983, Some observation of the flucuating radar signal. Preprints, 21st conf. on radar met., AMS, Boston, 167–172.

Fournier, A., D., Fussell, L. Carpenter, 1982, Computer rendering of stochastic models. *Comm. of the ACM*, 6, 371–384.

Frisch, U., P.L. Sulem, M. Nelkin, 1978, A simple dynamical model of intemittent fully developed turbulence. *J. Fluid Mech.*, 87, 719–724.

Goodchild, M.F., 1980, Fractals and the accuracy of geographical measures. *Math. Geol.*, 12, 85–98.

Goodchild, M.F., 1982, The fractional brownian process as a terrain simulation model.

Grassberger, P., 1983, Generalised dimensions of strange attractors. *Phys. Lett.*, 97, 227–230.

Hentschel, H.G.E., I. Procaccia, 1983, The infinite number of generalised dimensions of fractals and strange attractors. *Physica*, 8D, 435–444.

Kagan, Y.,Y., 1981a, Spatial distribution of earthquakes: the three-point moment function. *Geophys. R. astr. Soc.*, 67, 697–717.

Kagan, Y.,Y., 1981b, Spatial distribution of earthquakes, the four-point moment function. *Geophys. J.R. astr. Soc.*, 67, 719–733.

Kagan, Y.,Y., L., Knopoff 1980, Spatial distribution of earthquakes, the two-point correlation function. *Geophys. J.R. astr. Soc.*, 62, 303–320.

Kagan, Y.,Y., L. Knopoff 1981, Stochastic synthesis of earthquake catalogs. *J. Geophys. Res.*, 86, 2853–2862.

Kahane, J.P., 1974, Sur le modle de turbulence de Benoit Mandelbrot. *C.R. Acad. Sci. Paris*, A278, 621–623.

Korchak, J. , 1938, Deux types fondamentaux de distribution statistique. *Bull. de l'inst. inter. de stat.*, 3, 295–299.

Kolmogorov, A.N., 1941, Local structure of turbulence in an incompressible fluid for very large Reynolds numbers. *Comptes Rendues (Doklady) Academie des Sciencess de l'URSS (N.S.)*, 30, 299–303.

Kolmogorov, A.N., 1962, A refinement of previous hypotheses concerning the local structure of turbulence in a viscous incompressible fluid at high Reynolds number. *J. Fluid Mech.*, 13, 82–85.

Lerbel, F., 1985, The application of image synthesis to stereo mapping from radar. (in this book).

Lilly, D. K., 1983, Meso-scale variability of the atmosphere, in *Meso-scale meteorology theories, observations and models.*, D.K. Lilly, T. Gal-Chen Eds., D. Reidel, New York, 13–24.

Lilly, D., K., E. L., Petersen, 1983, Aircraft measurements of atmospheric kinetic energy spectra. *Tellus*, 35.

Lovejoy, S., 1981, A statistical analysis of rain areas in terms of fractals. Proc. 20th radar conf., AMS, Boston., 476–484.

Lovejoy, S., 1982, The area-perimeter relationship for rain and cloud areas. *Science*, 216, 185–187.

Lovejoy, S., J. Tardieu, G. Monceau 1983, Etude d'une situation frontale: analyse meteorologique et fractale. *La Météorologie*, 6, 111–118.

Lovejoy, S., B. Mandelbrot, 1985, Fractal properties of rain and a fractal model. *Tellus*, 37A, 209–232

Lovejoy, S., D. Schertzer, 1985a, Generalised scale invariance in the atmosphere and fractal models of rain. *Wat. Resour. Res.*, 21(8), 1233–1250

Lovejoy, S., D. Schertzer, 1985b, Scale invariance in climatological temperatures and the spectral plateau. *Ann. Geophys.*, 86(04), 401–409

Lovejoy, S., D. Schertzer, 1985c, Rainfronts, fractals and rainfall simulations. *Hydro. Appl. of Remote sensing and data trans.*, Proc. of the Hamburg symposium, IAHS publ. no. 145 (in press).

Lovejoy, S., D. Schertzer, 1985d, Scale invariance, symmetries, fractals and stochastic simulations of atmospheric phenomena. *Bulletin of the AMS*, (in press).

Mandelbrot, B., J. R. Wallis, 1968, Noah, Joseph, and operational hydrology. *Wat. Resour. Res.*, 4, 909–918.

Mandelbrot, B., JR. Wallis, 1969, Some long-run properties of geophysical records. *Wat. Resour. Res.*, 5, 321–340.

Mandelbrot, B. 1972, Possible refinement of the log-normal hypothesis concerning the distribution of energy dissipation in intermittent turbulence. in *Statistical models and turbulence*, M., Rosenblatt, C. Van Atta Eds., Lecture notes, physics, 12.

Mandelbrot, B., 1974, Intermittent turbulence in self-similar cascades: divergence of high moments and the dimension of the carrier. *J. Fluid Mech.*, 62, 331–350.

Mandelbrot B.B., 1975, Stochastic models for the earth's relief, the shape and fractal dimension of coastlines, and the number-area rule for islands. *Proc. Nat. Acad. Sci. USA*, 72, 2825–2828.

Mandelbrot, B., 1982, *The fractal geomtery of nature*. New York, Freeman and co., 461pp.

Mandelbrot, B., 1984a, Fractals in physics: squig clusters, diffusions, fractal measures and the unicity of fractal dimensionality. *J. Stat. Phys.*, 34, 895–930.

Mandelbrot, B., 1984b, Fractal sums of pulses and new random variables and functions (available from the author).

Mark, D. M., P.B. Aronson, 1984, Scale-dependent fractal dimensions of topographic surfaces: an empirical investigation, with applications in geomorphology and computer mapping. *Math. Geol.*, 16, 671–683.

Marshall, J.S., W. M. Palmer, 1948, The distribution of raindrops with size. *J. Met.*, 5, 165–169.

Mitchell, J., 1982, The nature of large scale turbulence in the Jovian atmosphere. Jan. 15, JPL publication 82–34.

Muller, J.P., 1985a, Nature of the atmospheric dynamics on Jupiter from the spectral distribution of cloud brightness, temperature and kinetic energy fields. *Geophys. Res. Lett.* (submitted).

Muller, J.P., 1985b, Applications of pattern recognition to geophysical fluid dynamical systems using remotely sensed data. IBM U.K. scientific centre report (in press).

Nastrom, G.P., K.,S., Gage, 1983, A first look at wave-number spectra fraom GASP data. *Tellus*, 35, 383–390.

Novikov, E. A., R. Stewart, 1964, Intermittency of turbulence and spectrum of fluctuations of energy dissipation. *Izv. Akad. Nauk. SSSR Ser. Geofiz.*, 3, 408–412.

Nguyen, P.T., J. Quinqueton, 1982, Space filling curves and texture analysis. Proc. 6th Int. conf. Patt. Rec., Munich, Germany, 282–285.

Nicolis, C., G. Nicolis, 1984, Is there a climate attractor? *Nature*, 311, 529–532.

Parisi, O., U. Frisch, 1985, A multifractal model of intermittency. *Proc. of the Varenna School*, Italian Phys. Şoc., M. Ghil Ed. (in press).

Peleg, S., J. Naor, R. Hartley, D. Avnir, 1984, Multiple resolution texture analysis and classification. *IEEE Trans. on PAMI.*, 6, 518–523.

Pentland, A. 1984, Fractal based description of natural scenes. *IEEE Trans. on PAMI*, 661–674.

Peyriere, J, 1974, Turbulence et dimension de Hausdorff. *C. R. Acad. Sci. Paris*, A278, 567–579.

Richardson, L.F. 1922, *Weather prediction by numerical process*. Republished by Dover, New York, 1965.

Rogers, R.R., 1971, The effect of variable target reflectivity on weather radar measurements. *Quart. J. Roy. Met. Soc.*, 97, 154–167.

Rothrock, D.A., A.S. Thorndike, 1983, Sea ice flow size distribution. *J. of Geophys. Res.*

Schertzer, D., S. Lovejoy, 1983, On the dimension of atmospheric motions, preprint vol., *IUTAM symp. on turbulence and chaotic phen. in fluids*, 141–144.

Schertzer, D., S. Lovejoy, 1985a, The dimension and intermittency of atmospheric dynamics. *Turbulent Shear Flow*, 4, 7–33, B. Launder Ed., Springer, New York.

Schertzer, D., S. Lovejoy, 1985b, Generalised scale invariance: symmetries, measures and dimension in anisotropic intermittent cascades. (submitted to *J. de Phys.*).

Schertzer, D., S. Lovejoy, 1985c, Generalised scale invariance in turbulence. *PCH PhysicoChemical Hydrodynamics Journal*, 6(5/6), 623–635

Takayasu, H. 1982, Differential fractal dimension of random walk and its applications to physical systems. *J. of the Phys. Soc. of Japan*, 51, 3057–3064.

Voss, R.F., 1983, Fourier synthesis of gaussian fractals: 1/f noises, landscapes and flakes. Preprints Siggraph conf., Detroit, 1–21.

Waymire, E., V. K. Gupta, 1981, The mathematical structure of rainfall representations, Parts 1–3. *Wat. Resour. Res.* 17, 1261–1294.

Waymire, E., 1985, Scaling limits and self-similarity in precipitation fields. *Wat. Resour. Res.*, (in press).

Yaglom, A.M., 1966, The influence of the fluctuations in energy dissipation on the shape of turbulence characteristics in the inertial interval. *Sov. Phys. Dokl.*, 2, 26–30.

11
Processing Satellite Infrared and Visible Imagery for Oceanographic Analyses

Paul E. La Violette
Naval Ocean Research and Development Activity
Bay St. Louis, Mississippi 39529-5004, USA

11.1. Introduction.

The physical events taking place in the ocean's surface layer often reflect the interaction of energy between the atmosphere and the ocean. Heating and cooling, evaporation and precipitation, convergence and divergence, upwelling and sinking, the very roughness of the surface—all are examples of this interaction.

In past studies of these phenomena, oceanographers collected large quantities of data on the surface and in the surface layer and used these data to derive a mean picture of the surface conditions of most oceans. However, the mean pictures of parameters such as sea surface temperatures or surface current fields do not show how the current or thermal patterns come to be, how they change, how they exist at any moment in time, or how they vary under certain oceanographic or meteorological conditions.

With the orbiting of the early Nimbus satellites in the late 1960's and the NOAA satellites in the 1970's, oceanographers were presented with a way to obtain a near instantaneous view of the thermal conditions of the ocean's surface. Figure 11.1 is an excellent example of these imagery. It shows the east coast of the United States and offshore temperatures associated with the meandering, complex path of the Gulf Stream. For the first time oceanographers were able to see the thermal distribution of this warm current in a manner far superior to that presented by Benjamin Franklin in the eighteenth century.

When given these repeated displays of the ocean's thermal radiation, oceanographers were faced with the dilemma of whether to believe that the presentations represented the true bulk temperature of the ocean or were mere manifestations of wind-driven surface effects. Many rationalized that since the displays were derived from the thermal radiance values of the first

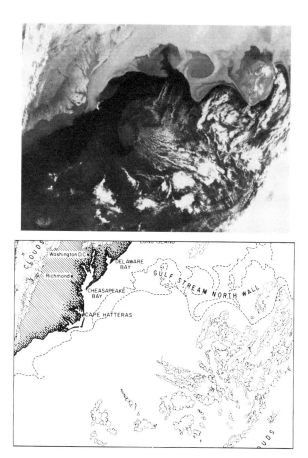

Figure 11.1. This enhanced, unregistered NOAA-5 AVHRR-IR (channel 4, i.e. 10.3-11.3 μm) image shows the eastern seaboard of the United States and the surface waters of a portion of the North Atlantic Ocean. The lighter areas of the picture are the colder radiated temperatures of the scene (whitest areas are clouds); the darker areas are the warmer temperatures. The image shows the surface thermal signature of the Gulf Stream as it forms a dynamic balance between the warmer waters of the Sargasso Sea and the cold slope waters off the continental shelf. This type of detailed display of the development of Gulf Stream meanders and cold and warm-core eddies (rings) is where satellite data excel in a manner that cannot be duplicated by in situ sensors.

molecular layer of the ocean, they did not represent the bulk temperature.

Others chose to believe that the data did represent ocean temperature conditions and sought to incorporate the data with their analyses of conventional ship data. These new users immediately encountered difficulties. The United States government agencies, NASA, and later NOAA, found that with the computers of that period, they were unable to provide the data to these users in a timely fashion. Six or eight months and even year-long waits were common. When the data were received, they were often still in the form shown in figure 11.1. Thus, because of the skewed cylindrical projection of the imagery, it was difficult for the ocean investigator to find accurately where his ship was in relation to the ocean features. Finally, whereas some idea of the relative strength of thermal gradients could be derived in the imagery, no actual values could be assigned to the gradients.

In recent years, with the advent of smaller and more powerful computers, and with the launching of more sophisticated satellite sensors, this situation has changed. Now ocean investigators can obtain satellite data in a more timely fashion, can place the data in suitable cartographic projections, and can derive true surface radiance values from the imagery. This Chapter will show the basic imagery manipulation techniques necessary to do these basic tasks so that satellite imagery can be processed in a manner suitable for oceanographic analysis.

Discussion will be limited to two proven spectral sources of satellite oceanographic information, the infrared and the visible. To ensure a proper understanding of these data, as used here, the physics of these spectral ranges, and the satellites and their sensors will be briefly discussed. Then the analysis techniques required to manipulate this type of information will be presented. Finally, examples of the analysis techniques used in actual ocean studies will be given.

11.2. Infrared and Visible Ocean Imagery.

Infrared.

Observations in selected infrared ranges can be used to determine equivalent blackbody temperatures emitted from the ocean surface. A number of problems arise, however, when these surface temperatures are measured from the altitude of a satellite. An understanding of these problems is needed before the data can be used in ocean studies. The notations used here are adapted from Holter (1970).

Planck's fundamental radiation equation gives the spectral distribution of radiation from a self-emitting perfect radiator (blackbody) with uniform temperature ($W_{B\lambda}$). Planck's law applies to the entire electromagnetic spectrum and can be expressed in terms of spectral radiant emittance as:

$$W_{B\lambda} = C_1 \lambda^{-5} \left[e^{(C_2/\lambda T)-1} \right]^{-1} \quad (11.1)$$

where C_1 and C_2 are radiation constants, λ is wavelength, and T is the temperature of the radiating surface. The units of the equation are w

$cm^{-2} \mu m^{-1}$. As most objects are not perfect radiators, the radiative efficiency factor (spectral emissivity), ϵ_λ, of most radiators varies with wavelength. This may be defined as

$$\epsilon_\lambda = W_\lambda / W_{B\lambda} \tag{11.2}$$

where W_λ is the spectral radiant emittance of an imperfect radiator (or greybody) for a given wavelength. Spectral emissivity of objects range from close to zero for poor radiators to unity for perfect radiators. Greybodies, such as the ocean, are nearly blackbodies within selected spectral bands. The important issue in most cases, therefore, is the characteristic emissivity of the specific band being monitored. Equations 11.1 and 11.2 can be used to obtain the spectral radiant emittance of any surface of known temperature and spectral emissivity. In addition to emitting energy, greybodies also reflect energy. Kirchhoff's law states that if a surface is optically opaque, then the spectral reflectivity, ρ_λ, and the spectral emissivity, ϵ, of the surface have the relationship

$$\rho_\lambda = 1 - \epsilon_\lambda \tag{11.3}$$

Thus, the infrared radiation from a surface is the result of a combination of (a) self-emission of the surface and (b) the reflection of a portion of the incident radiation on that surface. These principles are extremely important in any study of the ocean's radiant energy. They show that the infrared radiation emitted from the ocean surface comes from (a) the self-emission of the top-most radiating molecule of water and (b) the reflection of portions of incident energy from such atmospheric sources as clouds, water vapour, carbon dioxide, and aerosols. The sum of these two radiation components can be defined as the effective spectral radiance emittance of the ocean surface, $W_{e\lambda}$

$$W_{e\lambda} = \epsilon_\lambda W_{B\lambda} + \rho_\lambda W_{a\lambda} \tag{11.4}$$

where the self-emission component is $\epsilon_\lambda W_{B\lambda}$ and the surface reflection component is $\rho_\lambda W_{a\lambda}$. The additional effect of infrared absorption and emission by intervening atmospheric constituents (chiefly gas molecules of water vapour, carbon dioxide, nitrogen, and ozone) must be considered when the effective spectral radiance of the ocean surface is measured from satellite altitudes. From a satellite, the upward spectral radiance at a given zenith angle and wavelength, is given by:

$$N_\lambda(\theta) = W_{e\lambda}(T_S)\tau_\lambda a(\theta) + \int W_{a\lambda}(T_a) d\tau_\lambda(\theta) \tag{11.5}$$

where
$W_{e\lambda}$ is the effective spectral radiant emittance at temperature T_S,
$\tau_{\lambda a}$ is the spectral transmissivity through the atmosphere,

τ_λ is the spectral transmissivity through a layer that extends from a given altitude to the top of the atmosphere,

$W_{a\lambda}$ is the spectral radiant emittance of the atmosphere at temperature T_a.

In the above equation, the first term on the right is the radiant emittance to space of the ocean surface; the second term is the emittance of a cloud free atmosphere, integrated from the ocean surface to the top of the atmosphere.

In those spectral ranges where atmospheric gases are strongly molecularly absorbent, the atmosphere for all practical purposes is opaque. Fortunately, there are regions, or atmospheric "windows," in the spectrum where the absorption by these gases is weak (figure 11.2).

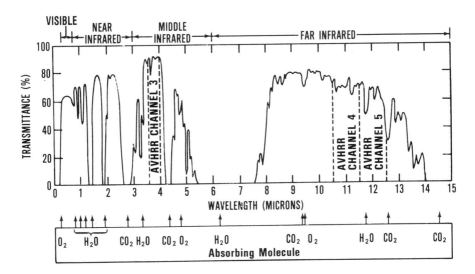

Figure 11.2. Characteristics of the electromagnetic spectrum significant to remote sensing of the ocean.

The boundaries of these windows are difficult to define because the transmissivity of the atmosphere depends on the concentration of the active absorbing gases present. In addition, the distance that radiation must travel through the atmosphere becomes a factor, since a low-angle look at an ocean region would pass through a thicker atmosphere and, hence, higher concentration of absorbing gases.

Planck's function gives the amount of energy radiated by a blackbody surface. From equation 11.1 and figure 11.2, it can be seen that the 3.4

to 4.2 and 8 to 12.5 micron window regions have the maximum emitted energy near 300 K. In the 3.4 to 4.2 micron window, reflected Sunlight and emitted radiation are almost equal in terms of energy. Thus, during daylight hours, data in this spectral range are contaminated by reflected Sunlight. As a result, the 8 to 12.5 micron range is the most effective for both day and night oceanographic use and is the one most used by passive ocean thermal sensors.

Visible.

The main signal-generating mechanism from the ocean in the visible portion of the spectrum (0.4 to 0.8 microns) is the backscatter of incident Sunlight. Thus, visible remote sensing is limited to day periods and, for practical purposes, to regions where the Sun is at least 30 degrees above the horizon. The ambient light from the ocean is backscattered in several ways: Fresnel reflection from the surface, and scattering and absorption within the water column being the more important.

Fresnel reflectance is the reflection of the Sun off the surface of the water (there is no penetration). Termed "Sun glitter," these reflections are generally so severe that no sub-surface colour information can be retrieved, and they reveal nothing about the biological/chemical constituents within the ocean. However, they do show spatial variations in the roughness patterns on the ocean surface. U.S. Space Shuttle Sun glitter photographs have been used qualitatively to provide information on ocean circulation in much the manner as imagery from synthetic and real aperture radars. Unfortunately, these photographic data have not been practically adapted to image processing. To show the oceanographic potential offered by these data (when such processing methods are developed) an example is included in the ocean studies portion of this paper.

The backscatter of light energy from below the surface (less spectral absorption) results in the upwelled spectral radiance (or colour) of the ocean. Measurements of upwelled radiance can provide information on the ocean's biological production and optical properties, and can act as tracers, or descriptors, of physical ocean events.

The depth at which the spectral radiance downwelled from the surface falls to $1/3$ of its initial energy is called "the first optical attenuation length" (Gordon and McCluny, 1975). Approximately 90 percent of the upwelled radiance is backscattered prior to this depth, and radiance from deeper in the water column is negligible. For practical purposes, then, upwelling spectral radiance of the ocean (measured at the water surface) may be considered to be the integration of radiant backscatter and the spectral absorption within the first optical attenuation length. An inverse relationship of the attenuation length: the diffuse attenuation coefficient, "k", is used as a measurement of ocean radiance. The k coefficient depends on the vertical distribution of scattering and absorption properties. However, it does not reveal the actual depth at which the optical variablity occurs. It is possible to have the same water-leaving radiance at the surface for two separate areas whose optical properties are vertically very different. For example, a relatively turbid homogeneous water mass can display similar surface radiance as a clear water mass, with a thin, turbid layer lying within the attenuation depth. Thus, any vertical variation in the optical properties that occurs in the first attenuation length can have a significant influence on the upwelling water-leaving radiation sensed by an ocean colour scanner.

The concentration of phytoplankton pigments are largely responsible for the ocean's optical properties. The spectral absorption ability of the pigments significantly affect upwelling radiance (Arnone and La Violette, 1986). The strong correlation between this biological pigment and water optical properties provides an excellent means of delineating the biological productivity of a region.

"Seeing" the ocean from space means looking at it through an atmosphere whose effects are significant at visible wavelengths; as much as 90 percent of the radiance detected could be due to atmospheric scattering (Gordon, 1978). Atmospheric scattering is caused by scattering from gas molecules (called Rayleigh scattering) and from aerosols or particulates (called Mie scattering). The molecular composition of the atmosphere is relatively constant so that the removal of the Rayleigh component of the Earth's radiance is not too complex. On the other hand, the aerosol effects are highly variable (spatially and temporally) depending on particle concentration, composition, size, and distribution. Methods of correcting for these effects will be discussed shortly.

Infrared and Visible Satellite Imagery Limitations.

It is important to remember that satellite imagery shows only one phenomenon: the distribution of the Earth's radiation energy. The imagery does not provide direct information on ocean currents or on the distribution of chemical or biological material. However, since the surface manifestations of currents are fairly delineated by the distribution of sea surface temperatures and colour boundaries, the surface location of such currents can be inferred by gradients depicted in satellite imagery of these spectral phenomena.

In a similar fashion, the abundance and distribution of many geo and bio-chemical properties are often indirectly related to the events that cause the variation in the distribution of thermal or selected colour gradients. In these cases a good knowledge of the relationship can make these data become highly useful for a geo or bio-chemical study. The essential point to remember is that while the infrared and visible data can be used to infer the occurrence and distribution of these events, they do not directly measure them.

The two most important limitations to the use of infrared and visible satellite imagery in oceanography is (a) the blockage by clouds of the ocean signal and (b) the integration of the ocean signal with that of the intervening atmospheric moisture and aerosols. Stripping the effects of the cloud cover and the atmosphere from the ocean signal is extremely difficult and requires extensive computer manipulation of the basic data. Several of the more succesful methods formulated to minimize these effects will be discussed shortly.

A further important limitation of infrared and visible satellite imagery is the shallow surface layer represented by the ocean signal. Indeed, the infrared imagery represents no depth at all, being the "skin" radiation of the ocean. These limitations become important in ocean regions undergoing extremely calm conditions and high solar radiation. In these areas, infrared imagery may display abnormally high temperature readings representative of extremely shallow surface heating, rather than the temperature of the first meter of the water column. When this type of surface heating occurs it is best to use night imagery of the region, since the oceans at night do not have solar heating to create stratified conditions and the shallow stratified layers left from the day are destroyed by convective mixing due to the ocean's normal thermal radiation. As most of the visible range imagery represent near-surface radiation (i.e. the portions of a visible image showing upwelled radiance, not Fresnel reflection), these data can be used to display frontal boundaries in lieu of the thermal imagery when shallow surface heating is a problem and night infrared imagery cannot be used.

Satellites That Can Provide Infrared and Visible Ocean Imagery.

No current satellite could be called an "oceanographic" satellite (in fact, there has only been one dedicated ocean satellite, the short-lived SEASAT). Although much satellite-gathered oceanographic information is available, it is obtained from satellite enviornmental sensors whose main purpose is not oceanography. Four satellite series can provide infrared and visible oceanographic data: GOES and METEOSAT (in geosynchronous orbits), and NOAA and Nimbus (in polar orbits). Because of the comparatively poor resolution of the GOES and METEOSAT data in comparison to that of the NOAA and Nimbus data, these discussions will be limited to the data from the latter.

The NOAA Advanced Very High Resolution Radiometer (AVHRR) is an oceanographically useful sensor that has been aboard all of the NOAA satellites since TIROS-N (i.e. all after NOAA 4). As the NOAA series of satellites are placed in Sun-synchronous orbits, the AVHRR aboard each satellite has the ability to detect repeatedly the thermal (AVHRR-IR) and visible (AVHRR-VIS) emissions of any region of the Earth at least twice each day (once during the dark side of the orbit and once during the Sun side). Thus, for cloud-free regions of the globe, 10-bit data from the AVHRR-IR are able to describe the surface thermal structure of the world's oceans approximately every 12 hours, with a nearly one kilometer resolution and an absolute accuracy of ± 0.6 degrees C. Detailed information on the NOAA satellites and the AVHRR are described in Schwalb, 1978 and Hussey, 1979.

In presenting NOAA-7 AVHRR-IR imagery (and usually all thermal imagery), the standard method is to display warm temperature as increasingly dark tones and cold temperatures as lighter tones, with the whitest feature usually being the cold tops of clouds. In some of the imagery presented here, a step function is used to display clouds as black. This highly useful display function de-emphasizes the seeming omnipresence of white clouds in an image, while allowing the presentation of sometimes subtle ocean thermal features.

Unlike the operational NOAA satellites, the sensors on the Nimbus research satellites are usually unique, with each new Nimbus spacecraft having a different suite of experimental sensors. The Nimbus-7 Coastal Zone Colour Scanner (CZCS), although proven to be a highly successful oceanographic sensor, is not scheduled to be placed on any of the forthcoming Nimbus spacecraft. However, much of what will be said here applies to other visible range sensors that may be placed in orbit in the future.

The 8-bit CZCS data has the capability of detecting subtle ocean colour changes from its five visible and near-infrared spectral bands, which were selected for their ocean viewing qualities. These six channels are centered at 0.443, 0.520, 0.670, 0.750 and 1.150 microns, with a resolution at nadir of 800 metres. The first four channels are in the visible and the fifth is in the near (reflective) infrared portions of the spectrum. The main purpose of the CZCS is to detect directly the variation and distribution of the ocean's chemical and biological material (figure 11.3). However, as figure 11.3 shows, it can also be used indirectly to provide information on currents and frontal features in the water. Although the polar orbit of Nimbus-7 is Sun-synchronous, the field of view of the CZCS limits the area of the Earth's surface that is viewed beneath the satellite. Rather than passing over the earth at the same time each day, the orbit is such that a given region at 40 degrees North or South is viewed approximately three out of four days at around local noon. A description of the Nimbus-7 satellite and CZCS can be found in Hovis *et al.*, (1980).

11.3. Satellite Image Analysis Techniques.

Basic Computer Techniques.

By their nature, satellite sensors produce vast amounts of data. In order to prepare these data for inclusion into oceanographic studies, computer analysis techniques must be used. Such techniques are best applied using computers interfaced to image processing and display hardware. In the study of a particular ocean region, a number of computer image analysis techniques can be used. However, included in that number should be one or more of four basic techniques: selective enhancement, geographic registration, absolute ocean radiation and multiple-image composition. In addition, combinations and subsets of these basic techniques, such as edge enhancement or sequential displays (loop movies) of registered images, can be used to detect and measure motion. This portion of the Chapter will describe these four basic techniques.

Selective Enhancement.

Of the four basic image analysis techniques outlined above, one of the most useful and easiest to apply is selective enhancement. Although the grey-scale wedge in the imagery is limited to a finite number of grey bands, these bands can be optimally assigned in order to enhance selected ocean phenomenon (figure 11.4a).

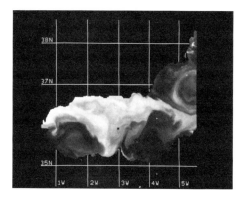

Figure 11.4a. A NOAA-7 AVHRR Channel 4 image of the Alboran Sea at near 1500 hrs local on May 12 1986. The grey tones in the image have been linearly enhanced to show the ocean features.

When the grey (or colour) shading is made equivalent to absolute temperature or bio-optical values, the enhancement technique can be useful quantitatively, as well as qualitatively. Another method to emphasize the image's oceanic information utilizes edge enhancements or perspectives, such as shown and described in figure 11.4b and 11.4c.

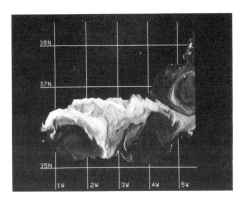

Figure 11.4b. This edge enhancement (using the same data as figure 11.4a) has been processed with a modified Frei-Chen edge enhancement filter. Filter weights were chosen so that the intensity at each point was replaced by a 3 × 3 finite difference approximation to the original image. Linear contrast enhancement was applied to the image prior to the filtering. The results provide the impression of relief based on the sea surface temperatures.

Although these methods are not quantitative, they do present a method of presentation that, by emphasizing certain aspects of an image, can be highly informative.

The problem with linear assignments of grey-scale wedges in an image is that if large thermal differences are present in a specific scene, the depiction of small, but important, thermal changes may be overpowered. Figure 11.4a is a good example of this. The grey-scale shadings for the small thermal gradients are still present in this image, but the changes are too small for the eye to detect. Stepped or logarithmetic grey-scales can be used to overcome this difficulty. However, this calls for visual interpretation by the user and may interfere with the flow of information. Colour wedges can be used instead of the linear grey wedge to bring out these small gradiant features. An example is shown in figure 11.5 (see colour plates), which presents the same scene as figure 11.4, but in colour.

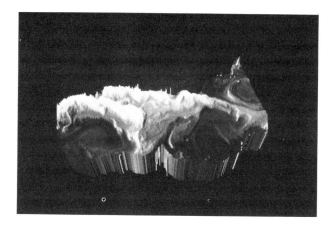

Figure 11.4c. A pseudo three-dimensional enhancement (again using the same data as figure 11.4a) in which the radiation values were used to provide linear enhancement, as well as the dimension of height; warm values are higher and cold values are lower.

Geographic Registration.

Perhaps the most important aspect of computer processing of satellite imagery is the accurate geographic location of the ocean features seen in the imagery. Accurate geographic registration allows these features to be correlated with data from other satellites, or from ships or aircraft. Computer compatible tapes (CCT's) of NOAA satellite data distributed by many of the various national satellite data distribution centers allow accurate geographic registration to be done. These have geographic positions incorporated in a systematic fashion throughout the AVHRR data in order to produce imagery with location accuracies to standard deviations of 1.7 km about mean errors of 3.7 km (Clark and La Violette, 1981). The accuracy may be further improved if the ocean region has well-distributed land features.

This ability to locate the data accurately can be used to overcome a major problem with the unprocessed satellite data. As a result of the sensor scanning mirror and the forward motion of the satellite, the raw satellite data forms an image whose projection is a skewed rectangle. Since the satellite's orbit processes slightly each day, there is no common Earth projection from one day's image to the next. Standard map projection, such as Mercator or polar stereographic, can be made by incorporating the geographic locations in the AVHRR data into two-dimensional, third-order polynomials. Thus, sets of NOAA imagery can be constructed that have a common projection. Such sets permit studies that involve measurements on temporal and spatial scales that were not possible using the unwarpped imagery.

Figures 11.4a and 11.5 are examples of an image that has been warped to Mercator projections with sufficient accuracy (±1.0km) to permit the overlay of other data. In the case of figure 11.4a, the succesive imagery was used to produce a time-lapse sequence or movie loop of ocean features covering a 10-day period with no noticeable mis-registration jitter. An example of such a series (for the same area but another year) is shown in figure 11.6.

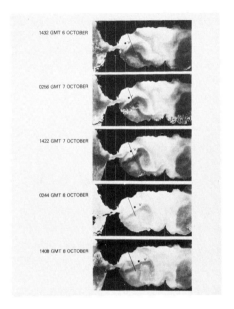

Figure 11.6. Mercator-registered NOAA 7 AVHRR-IR imagery of the Alboran Sea collected as part of the Donde Va experiment during October 1982. The dot and the V in the imagery show the advection of two sub-mesoscale cold features about the permanent gyre located in the Alboran Sea (the line refers to the position of a set of current moorings). A close examination of the imagery shows that other unmarked features were also being advected and that their generation near Gibraltar was probably tide-related. Continuous monitoring of the displacement of the cold features in successive images shows their apparent origin east of Gibraltar, their average speeds of 40 cm/s around the gyre, and their apparent entrainment into the incoming Atlantic water east of Gibraltar. (La Violette, 1984).

An example of the overlay of *in situ* data on accurately registered NOAA thermal imagery will be shown in the last section of this paper.

Absolute Ocean Radiation.

One of the most pressing prerequisites for the quantitative use the satellite data is the accurate determination of the absolute ocean radiation. In view of this, it is unfortunate that this is the most difficult of the analysis techniques. The technique necessitates, first, the radiometric calibration of the satellite sensor, and second, the removal of that portion of the signal that is the result of the atmosphere. The remarks that follow will highlight some of the problems in applying this technique and, in doing so, will emphasize some of the basic limitations of applying satellite data to the study of ocean processes.

Infrared Atmospheric Correction.

Radiometric calibration of the NOAA AVHRR-IR channels is carried out continuously onboard the satellite. If one assumes the output of the thermal detector is linear with input energy, a calibration curve can be established using only two points. One point is the scanner view of deep space (zero radiance) and the other point is the view of the scanner housing (290 K). Four platinum resistance thermistors, whose outputs are included in the data stream of each scan line, monitor the exact temperature of the housing. This data is averaged over several sweeps of the sensor mirror and then transmitted with the terrestrial radiation data to the ground station. The resultant curve provides a dynamic calibration, which is used to track and to compensate for any drift in the performance of the detector and associated electronics. CCT's received from central satellite data collection stations, such as NOAA National Environmental Satellite Data Information Service (NESDIS), normally contain these calibration data.

With infrared imagery, the problem of atmospheric attenuation of the ocean signal is especially prevalent in data taken at mid- and low-latitudes in summer. Of the several possible solutions to this problem, the one most often used is the multichannel single-satellite approach (McClain *et al.*, 1977).

The three NOAA AVHRR-IR channels, centered at the 3.7, 11, and 12 micron atmospheric windows, have relatively high transmittances with regard to middle and far-infrared spectral range energy emitted by the ocean surface. The most significant atmospheric absorption constituents in these regions are water vapour and aerosols. However, the amount of absorption varies for each spectral region. The 3.7-micron channel has a transmittance of 90%, the 11-micron channel has a transmittance of 80%, and the 12-micron channel transmittance is 75%. Although limited to night use, one of the preferred channels to use because of its high transmittance rate is the 3.7-micron channel; however, high noise levels in this channel in all of the post-TIROS-N AVHRR-IRs have precluded any real use of a three-channel algorithm. To date, only a two-channel algorithm has been used. These variations in infrared transmittance for the same ocean scene forms the rationale for the multiple-channel algorithm. The algorithm utilizes a correction made by subtracting the effects of one channel from another with the remainder being a factor of the amount of moisture present in the atmosphere. The two-channel equation takes the form

$$\text{MCSST} = A_1(T_4) + A_2(T_5) + A_3 \tag{11.6}$$

where

MCSST is the multichannel sea surface temperature,

T_4 is the measured channel 4 radiative temperature,

T_5 is the measured channel 5 raditive temperature, A_1, A_2, and A_3 coeffients derived from a best-fit to empirical surface data.

MCSST imagery are produced on a routine basis using the above algorithm. Such imagery is of particularly high quality in high-thermal-gradient ocean regions. However, it has been noted that MCSST imagery of lowthermal-gradient regions appears noisy. La Violette and Holyer (1987) propose that the random noise in AVHRR channels 4 and 5 is considerably amplified in the MCSST process. The noisy MCSST imagery that result appear particularly poor in areas where the amplified noise in is approximately the same magnitude as the SST gradients. That is, although the increase in noise is always present, it is more noticeable in imagery of thermally-quiet areas.

The southern portion of the western Mediterranean Sea in winter, where the average temperature range is less than 1 degree C, is an example of a region where the MCSST signal-to-noise ratio will be about 1. This can be compared to the MCSST signal-to-noise ratio of approximately 10 for the Gulf Stream in winter, and where the SST variations range over 10 degrees C. An example of the difference between an MCSST image and the channel 4 image of the Mediterranean is shown in figure 11.7.

This is not to say that MCSST noise is not present in high thermal-gradient MCSST images, but that it is seldom noticeable because it is small with respect to the magnitude of the SST gradients. It is not surprising, therefore, that with only a 1.0 degree C SST range, the MCSST image would look noisy, as the noise is of the same order of magnitude as the signal. This also explains the sharp difference between the channel 4 and the MCSST image of a low thermal gradient region.

The above remarks show why visually appealing MCSST imagery should not be expected in areas of low thermal contrast. The imagery can be both esthetically and numerically improved by running some type of smoothing function over the image. The averaging will tend to reduce the random noise and thus present a more accurate thermal value for a given pixel. Of course, the smoothing will also blur the details of the thermal

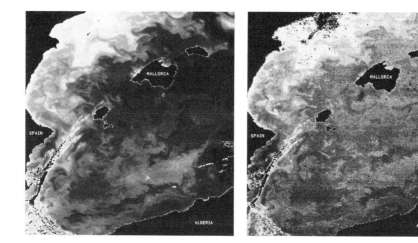

Figure 11.7. NOAA-7 AVHRR data for the western Mediterranean Sea on December 21, 1986, shown as a channel 4 image (a) and an MCSST image (b). Both images have been enhanced to bring out their oceanographic characteristics.

features, with the degree of blurring depending on the amount of smoothing. Some middle course must be decided on by the investigator that results in the data presentation that best fits the requirements of his or her study.

It would appear, then, that for studies of thermally-quiet ocean regions that require actual temperature values, an MCSST image with some smoothing is needed. However, if displays of the thermal variations of these same areas are the only requirements of a study, it would appear better to use the channel 4 or 5 imagery. For studies utilizing movie loops, which require a consistency in the grey tones assignments from frame to farme, a combination of channel 4 and MCSST could be used, with the MCSST values dictating the assignment of grey-tone values in the detailed channel 4 imagery.

A final word on thermal gradients. There must be some gradient present in either channel 4 or 5 for the multichannel algorithm to show a front (i.e., if the thermal field for both channels is flat due to very large water vapour attenuation, then no set of coefficients can bring out the true SST field). Thus, in some tropical environments, infrared SSTs cannot see the actual SST gradients.

Visible Atmospheric Correction.

The greatest atmospheric problem contained in visible data of the ocean is not that of atmospheric gases but, rather, that of aerosols. Aerosols are not only the major absorber and scatterer of the visible atmospheric signal, but their effect is far greater than with infrared data, with the loss of the ocean signal amounting to as much as 90%.

Correcting for the effects of aerosols in CZCS data requires extensive computer manipulation involving assumptions about the scattering properties of the local atmosphere and ocean. The most widely accepted method uses a mathematical model of the atmosphere based on single-scattering theory (Gordon and Clark, 1980a):

$$L_W(\lambda) = L(\lambda) - L_R(\lambda) - \frac{F(\lambda)}{F(670)}[L(670) - L_R(670)] \qquad (11.7)$$

where
$L_W(\lambda)$ is the upwelling radiance from the ocean at wavelength λ,
$L(\lambda)$ is the observed radiance from space,
$L_R(\lambda)$ is the calculated Rayleigh scattering radiance of the atmosphere,
$F(\lambda)$ is solar flux.

This method assumes that the CZCS 0.670-micron channel can be used as a measure of the atmospheric aerosol concentration. Through a weighted subtraction of this channel with appropriate subtraction of Rayleigh scattering, the effective upwelling radiance from the ocean surface can be computed for the 0.443, 0.520, and 0.550-micron channels of the CZCS. However there are several problems associated with this correction method.

The first problem involves the proper selection of the Angstrom exponent. The Angstrom exponent relates the wavelength dependence of the aerosol optical thickness to the single scattering albedo of the aerosols. Gordon and Clark (1981) suggest that, in any CZCS image, the radiance values of clear water be used to obtain this value. However, this technique requires knowing where clear water areas are located within the CZCS image. Furthermore, the assumption is made that the Angstrom exponent over clear water is the same throughout the image. If this assumption is not valid, then the selected Angstrom exponent value for one part of the image will produce erroneous results when applied to other portions of the image (Arnone, 1983). Arnone and La Violette (1986) presented a method to overcome these difficulties. This method uses an interactive analysis procedure to define the optimum coefficient for the entire image scene. The method is involved and its description is too lengthy for inclusion here. The reader is advised to consult the paper for a complete description.

A second problem with the Gordon and Clark (1981) correction procedure arises from the assumption that water has zero upwelling radiance at 0.670 microns. This is logical for open-ocean areas containing minimal concentrations of suspended sediments. However, significant upwelled radiance at 0.670 microns does occur in turbid coastal waters (Smith and Wilson, 1981; Arnone, 1983a,b). Use of the correction procedure in these coastal areas will result in an underestimation of the absolute upwelling radiance in the corrected CZCS channels data.

Arnone and La Violette (1986) assumed an inherent relationship between "k" and phytoplankton concentration, and used the ratio of the absolute upwelling radiance for the 0.443 and 0.550-micron channels (i.e. 0.443/0.550) to derive both the diffuse attenuation coefficient "k" at 0.490 microns and the phytoplankton concentration.

Multiple-Image Composition.

Clouds in an infrared satellite image of the ocean are a constant problem. This may be partially overcome using a method that takes advantage of the fact that ocean events change slowly compared to atmospheric events. The method composites the highest temperature values, pixel by pixel, for a given region using infrared data collected over a period of several days (La Violette and Chabot, 1969). The technique is applicable to any set of accurately registered thermal IR imagery using a common projection (figure 11.8).

This method has limitations. Short-term ocean events are smeared in the final composite, and the composite is biased toward the period when the area was cloud-free. A user must take these limitations into account and weigh them according to the desired objective. The problem is one of using the least amount of satellite passes that will give the most cloud-free conditions, and doing the least smearing of ocean features.

Ocean data from the higher latitudes allow the most advantageous use of this technique. In these regions, the increased number of overflights each day by the polar-orbited NOAA and Nimbus satellites allows composites to be made of several passes in a given day. As many of the storm systems in the polar regions are fast moving, the chances are increased of a satellite passing over a cloud-free period during a given day. Compositing all the passes that occur in one or two days can minimize the amount of ocean changes that may have taken place.

Using a sliding composite technique Holyer *et al.*, (1980) expanded the compositing method developed by La Violette and Chabot (1969). A selected number of days are initially composited to form the first of a series of images. The next composite image uses the same number of days, but drops the first day and adds the next chronological day to the series. The resulting composite image is, thus, one day removed from the first composite image. The method is continued until a satisfactory series is made that can be made into a loop movie. One potential application of

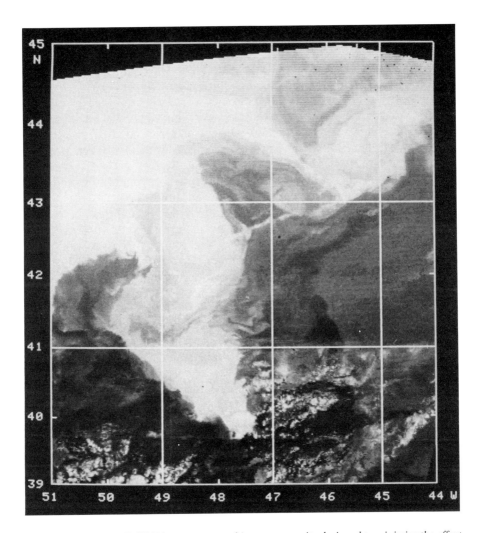

Figure 11.8. Four TIROS-N imagery merged into a composite designed to minimize the effect of clouds. The composite was formed by comparing the temperatures for four days at each pixel location and retaining at each address the warmest temperature for those fours days. (Holyer et al., 1980.)

this technique is to make loop movies of satellite imagery to compare with movie loops derived from numerical models.

Each of the atmospheric correction methods described had its own problems and limitations. However, the methods have been proven to be useful, and the discriminating investigator will pick and choose from them, using a matrix of methods that will most adequately satisfy his or her particular needs.

11.4. Examples of Processed Satellite Imagery: Alone and With Conventional Data Oceanographic Analysis.

Examination of the two processed images in figure 11.9 is interesting in that it shows how well colour and thermal data can be used to describe both physical and biological ocean events. These pictures, taken from an unpublished study by Mark Abbott and Phil Zion of the Jet Propulsion Laboratory show the California Current in its southern migration along the California coast interacting with wind-induced upwelling.

Particularly striking in both images is the regular spacing of the coastal filaments that stretch 100 to 300 km offshore. Historical data had indicated a much smoother flow relationship along the coast, with what was (then) thought to be short-term anomalous changes. However, over the last decade satellite data has indicated the frontal structure to be not only more varied, but also has a regular variation that is obviously dependent on the basic mechanisms creating the regional flow.

Abbott and Zion indicated that, in these summer images, prevailing winds from the north west have driven the coastal surface waters offshore, inducing an upwelling of cooler sub-surface water. A comparison of the two images shows that the cool, sub-surface water upwell along the coast is rich in nutrients. As these move seaward in the Sunlit surface layer, they provide an ideal environment for phytoplankton growth. These and other satellite data have shown that much of the phytoplankton growth in the coastal zone is entrained into jets or filaments and carried far offshore (8 and 9 in the images).

Abbott and Zion believe that this pattern of distribution has an important biological implication: Since phytoplankton levels are about 100 times higher along the coast than in the nutrient-poor offshore waters, this transport of phytoplankton is the most likely food source for the high populations of zooplankton observed several hundred kilometers from the coast.

The Nimbus-7 CZCS image showing Baja California (figure 11.3) lies only slightly further south than figure 11.9. It also shows the effect of regional upwelling on the phytoplankton concentrations (1 and 2 in the image). In the south, the Costa Rica Current (3) flows north, bringing warm water from equatorial regions to meet the cooler southward-flowing California Current. Near the southern tip of Baja, a cloud/fog bank (the white area) has formed over the cold water at an upwelling site.

Dennis Clark of NOAA, in an unpublished study of these and numerous other CZCS images of the region, has shown how conditions in the Gulf of California are quite different from those lying just off the coast in the

Pacific. Low-nutrient surface water flowing into the Gulf from the south retains generally low levels of phytoplankton (4), except for a narrow area of upwelling along the coast of mainland Mexico. North of the Midriff Islands (5) strong tidal currents cause intense mixing as they pass over shallow sub-surface ridges, bringing deeper nutrient-rich waters to the surface. The result is an abundance of phytoplankton, indicated by reds in the image.

Clark has processed the four smaller, composite mean CZCS images (included in figure 11.3) to illustrate the marked seasonal variations in the distribution and abundance of phytoplankton in this region. They were derived from 115 CZCS images, with each image containing clear sky conditions to allow the depiction of pigment concentration. Fom left to right, the images show winter (December 1978 to March 1979), spring (April to May 1979), summer (June to September 1979), and autumn (October to November 1979). Transient features, such as the filaments seen in the upper image, have been smeared out in the averaging process required to produce the mean picture and are not apparent in the seasonal means.

Except for the summer, the southern half of the Gulf has a reasonably high concentration of phytoplankton throughout the year. The most striking seasonal change in the Gulf is from low phytoplankton abundance in the summer to high phytoplankton abundance in the autumn. This increase is due to upwelling induced by favourable northwest winds in the autumn; this also results in high abundance on the Pacific side of Baja. These patterns, very difficult—if not impossible—to measure from scattered ship observations, represent the first statistically valid analysis of the seasonal variation of phytoplankton distributions. The data upon which this one-year time series is based far exceeds all at-sea measurements of phytoplankton ever made in this area.

The western Mediterranean Sea has been the object of several recent international investigations that relied heavily on satellite data analysis techniques: Donde Va (Donde Va Group 1984); the Gibraltar Experiment (Kinder and Bryden, 1986); and the Western Mediterranean Circulation Experiment (WMCE) (La Violette, 1987). The following discussions are taken from studies conducted as part of these investigations. They show how interactively processed satellite imagery can be combined with conventional ocean data of an ocean region.

Briefly, the Mediterranean is an evaporative, semi-enclosed sea whose only substantial connection to the world ocean is a two-layer system of flow in the Strait of Gibraltar. Atlantic water flowing into the Mediterranean Sea at the surface in the Strait overrides a deeper layer of dense Mediterranean water outpouring into the Atlantic Ocean. The surface flow of the fresher Atlantic water replaces both the water evaporated within the sea and the deep outflow of the highly saline Mediterranean water. The two Alboran Sea basins are the first of the Mediterranean basins encountered by the replacement Atlantic water and within them occurs most of the mixing of the fresher Atlantic water with the highly saline Mediterranean water occurs within them.

Although the regional circulation of the western Mediterranean Sea is mainly salinity-driven, proper interpretation of infrared and visible satellite imagery can indicate the eastward migration of the constantly modified Atlantic water. The satellite thermal and visible data indicate that the mean surface flow pattern in the Alboran Sea is that of two adjacent anticyclonic gyres, called the Western and Eastern Alboran Gyres (figure 11.4). In addition, preliminary WMCE studies indicate that the attitude the Eastern Alboran Gyre makes with the African coast may determine the configuration (meandering as opposed to coastal eddies) of the current found further east along the African coast.

The input of Atlantic water into the Alboron Sea is greatly influenced by oceanic events that occur in the Strait of Gibraltar. La Violette and Lacombe (1987) show evidence that the flow within the Strait does not move as continuous currents but as tidal-induced pulses. Their study indicates that during each tidal cycle a pulse of Atlantic water is emitted from the Strait of Gibraltar into the Alboran Sea; the strength of the pulse depends on the phase of the tide.

Figure 11.6 shows a series of sequential infrared satellite images that illustrate this pulsed flow. This series is extracted from a 10-day loop movie of the cold-water mesoscale features that periodically form at the eastern end of the strait. La Violette and Lacombe (1987) suggest that the mesoscale features are formed of cold water that upwells in the Strait during the comparatively slack period between the tidal-induced pulses of warm Atlantic water. Upon entering the Alboran Sea, the features become entrained in the rotation of the gyre and circle the sea.

Figure 11.10 is part of a study of several of these features' migration about the gyre (La Violette, 1984).

A comparison with wind conditions occurring at the time indicated no apparent effect on the movement of these features. The passage of the features over current moorings, and the dropping of Airborne Expendable Bathythermographs (AXBTs) and sonobuoy drifters, indicates that the features have strong flows associated with them and that they extend approximately 75m into the surface layer (Figures 11.11 and 11.12). More extensive studies of these data are still taking place. However, their presentation here is to show how the different types of *in situ* data can be merged with the satellite data to determine the oceanographic events taking place over an area.

Although the anticyclonic circulation of the Eastern Alboran Gyre has been noted in satellite infrared imagery for several years (e.g. Phillipe and Harang, 1982), the limited *in situ* investigations made prior to 1986 had not indicated a second gyre in the area (e.g. Laniox, 1974). Modelling studies of the sea (probably influenced by the earlier studies) also do not include

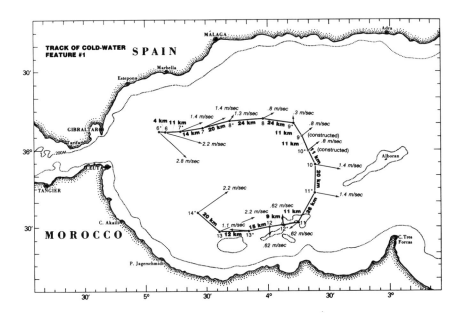

Figure 11.10. The complete track about the Western Alboran Gyre of the cold-water feature shown in figure 11.6. The dates in October are shown beside the location points, and the asterisk represents the morning position. The arrows represent wind speed and direction close to the time of each plotted position.

the feature (e.g. Preller and Hurlburt, 1982). If anything, these studies show the eastern Alboran Sea to have a semicyclonic circulation (figure 11.13).

In addition to the examination of infrared imagery, preparations for WMCE included a field study of the Almeria-Oran front during Space Shuttle Mission STS-41-G. Working in unison with the Shuttle, three aircraft flights were made over the area during the period 6–13 October 1984. AXBTs were dropped during these flights in an attempt to obtain a thermal cross-section of the front simultaneous to photographs being taken from the Shuttle.

A mosaic of three Shuttle photographs taken of the area at five-second intervals at approximately 1 hour after local noon on 8 October 1984, is shown in figure 11.14. The geographic location of the photographs in relation to the thermal front is shown in the figure by a NOAA infrared image taken on the same day, but approximately 3 hours later.

The photographs show the Sun reflecting off the water's surface (although some clouds are present in the NOAA image, little, if any, of the Shuttle photograph contains clouds). Since the surface of the ocean is not

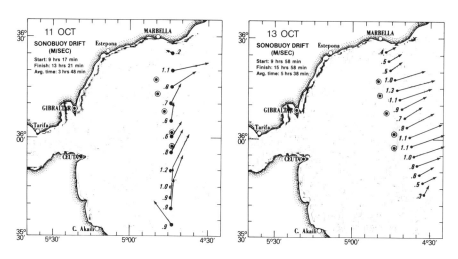

Figure 11.11. Smoothed MCSST values (grey isolines and numbers) taken from NOAA-7 AVHRR data for October 11 and 13, 1982, with current mooring positions (circled dots) and sonobuoy drift vectors superimposed. This type of accurate superpositioning of in situ data with satellite data is necessary in order to conduct an integrated analysis of the various data.

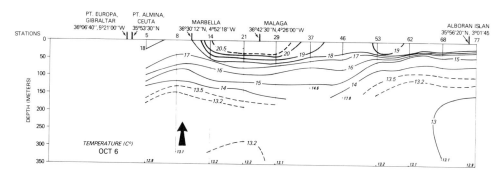

Figure 11.12. Airborne XBTs dropped in an east-west line across the Alboran Sea for 6 October showing the subsurface vertical extent of one of the features shown in the satellite images in figure 11.6.

smooth, the Sun does not appear in its reflection as a disc, but as a distorted image whose distortion is determined by the amount of surface roughness and the solar incident angle. In simple terms, the ocean's roughness varies according to the pressure of the wind, air/sea-surface temperature differ-

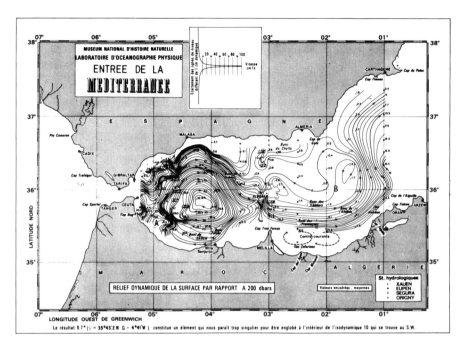

Figure 11.13. An analysis of ship data by Lanoix (1974) showing the dynamic topography based on 200 dbar. Although the Western Alboran Gyre is well defined in this analysis, the Eastern Alboran Gyre is not present.

ences, and water movement. Because of this, the patterns created by the ocean roughness can delineate ocean events involving these parameters (The effect is similar to the displays of imaging radars, such as Synthetic Aperture Radar (SAR); although the solar reflectance in the photographs does not involve coherent energy).

To aid with the comparison of the Shuttle photographs, the aircraft was equipped with photographic and television cameras, an infrared scanner (uncalibrated), and a search radar. All showed manifestations of the Almeria-Oran Front, including the aircraft search radar. The AXBTs and other measurements showed a sharp decrease of approximately 2 degrees C in the region of the front, and a drop and a weakening of the thermocline on the western side of the front.

Based solely on the persistence of the feature in the satellite imagery and the results of the Space Shuttle aircraft survey, several oceanographic cruises were planned and conducted in the region of the Almeria-Oran front during the WMCE field year.

Figure 11.14a. A mosiac of three photographs taken within seconds of one another from aboard U.S. Space Shuttle Mission STS-41-G near local noon on 8 October 1984 from an altitude of approximately 200 km. In each of the almost cloud-free photographs, the Sun's reflection on the ocean surface is shown, with its position determined by the angle of the Sun and the spacecraft in relation to the ocean surface. As the shuttle moves, the angle changes and the reflection moves. Thus, as the shuttle sweeps over the ocean, the reflection moves as a spotlight, illuminating the vast interconnection of ocean features. In this mosiac, stress lines related to the shear of the currents are prominently defined. Also seen are anticyclonic spiral eddies associated with the front. The prominent east-west lines are ship tracks. At this time, attempts to digitize and process these type of photographs by computer in a fashion that will allow the manipulation of the subtle ocean features have been unsuccessful. It is presented here partially to indicate possible directions for future analytical technique development.

11.5. Summary.

This Chapter has shown that certain basic image analysis techniques must be applied to satellite data for it to be a reliable and quantitative source of oceanographic information. It has stressed that proper registration and atmospheric correction are the most important of these basic techniques (especially with visible range data where up to 90 percent of the signal may be due to the intervention of the atmosphere). It has demonstrated that registration to a set graphic projection can be extremely useful in al-

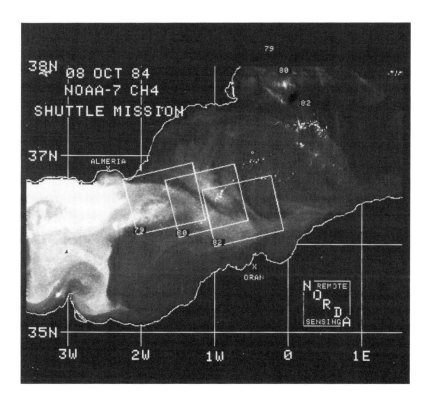

Figure 11.14b. A NOAA-7 infrared image taken approximately 3 hours after the Shuttle's passage, is presented with the geographic location of the mosiac. It is difficult to place marks on the mosiac without interfering with the visual details of the photographs. The reader is therefore asked to compare by eye the infrared image and the mosiac and to note the extreme wealth of similarities between the thermal features and the frontal features, including the lines of shear and eddy fields. (NASA photographs 38-079, –080 and –082. Satellite data collected by Royal Aircraft Establishment, Farnborough, England. Image processing by NORDA IDSIPS. NASA photographs 38-079, –080 and –082. From La Violette et al., 1986) Farnborough, England. Image processing by NORDA IDSIPS.

lowing several days of data to be studied as a temporal, as well as spatial, continuity. However, it has also shown that when the data are accurately calibrated and atmospherically corrected, the usefulness of the data can be expanded considerably when co-registered with other data sets and used as part of an overall analysis of the data available for a study region.

Emphasis has also been placed on specific concepts in the use of the data for oceanographic purposes: that of understanding the oceanography of the region being studied and knowing the limitations of the satellite data.

These areas must be understood before the computer-processed imagery can be properly exploited.

With these understandings and with the utilization of the basic image analysis techniques, satellite imagery can provide an imaginative oceanographer with a powerful analytical tool.

References.

Arnone, R. A., 1983a, Evaluation of CZCS and LANDSAT for coastal optical and water properties, *17th International Symposium on Remote Sensing of Environment*, Ann Arbor, MI.

Arnone, R. A., 1983b, Water Optics of the Mississippi Sound, *Naval Ocean Research and Development Activity (NORDA) Report 63*.

Arnone, R. A., and P. E. La Violette, 1984, A method of selecting optimal Angstrom coefficients to obtain quantitative ocean color data from Nimbus 7 CZCS, *SPIE* Vol 489 (Ocean Optics VII), p 187–194.

Arnone R. A., and P. E. La Violette, 1986, Bio- and chem-optical variability of the shear eddies associated with the African Current off Algeria as assessed by the Nimbus-7 Coastal Zone Color Scanner, *JGR*, v 91, no C2, p 2351–2364.

Austin, R. W., and T. J. Petzold, 1980, The determination of the diffuse attenuation of sea water using the Coastal Zone Color Scanner, in *Oceanography From Space*, edited by J F R Gower, p239

Clark R., and P. E. La Violette, 1981, Detecting the movement of oceanic fronts using registered TIROS-N imagery, *Geophys Res Lett*, v 8, no 3, p 229–232.

Donde Va Group, 1984, Donde Va: An oceanographic experiment in the Alboran Sea, *EOS*, v 65, no 36, p 682-683.

Gordon, H. R., 1978, Removal of atmospheric effects from satellite imagery of the oceans, *Applied Optics*, v 17, p 1631–1636.

Gordon, H. R., and V. W. Brown, O. B. Brown, R. H. Evans, and D. K. Clark, 1983, Nimbus-7 CZCS: reduction of its radiometric sensibility with time, *Appl Opt*, v 22, p 3929.

Gordon, H. R., and D. K. Clark, 1980a, Atmospheric effects in remote sensing of phytoplankton pigments, *Bound Lay Met*, v 18, p 299–313. Gordon, H. R., and D. K. Clark, 1980b, Remote Sensing optical properties of a stratified ocean: an improved interpretation, *Appl Opt*, v 19, 3428. Gordon, H. R., and D. K. Clark, 1981, Clear water radiances for atmospheric correction of coastal zone color scanner imagery, *Appl Opt*, v 20, no 24, p. 4175–4180.

Gordon, H. R., and W. R. McCluney, 1975, Estimation of the depth of sunlight penetration in the sea in remote sensing, Appl Opt, v 14, n 2, p 413-46.

Gordon, H. R., J. L. Mueller, S. F. El-Sayed, B. Sturm, R. C. Wrigley and C. Yentsch, 1980, Nimbus-7 Coastal Zone Color Scannner: system description and initial imagery, Science, v 210 (4465) p 60–63.

Holter, M., 1970, Chapter 3. Imaging with non-photographic sensors, in *Remote Sensing With Special Refrence to Agriculture and Forestry*, National Academy of Science, Washington, DC

Holyer R. J., P. E. La Violette, and J. R. Clark, 1980, Satellite oceanography research, development, and technology transfer, *Proceedings of Fourteenth In-*

ternational Symposium on Remote Sensing of Environment, 23–30 April 1980, Costa Rica.
Hovis, W. A., D. K. Clark, F. Anderson, R. W. Austin, W. A. Wilson, E. I. Butler, D. Ball, H. R. Gordon, J. L. Mueller, S. F. El-Sayed, B. Sturm, R. C. Wrigley and C. Yentsch, 1980, Nimbus-7 Coastal Zone Color Scanner: system description and initial imagery, *Science*, v 210 (4465) p 60–63.
Hussey, W. J., 1979, The TIROS-N/NOAA Operational Satellite System, *NOAA NESS Tech. Memo.* 95, National Environmental Satellite Service/NOAA, Washington, D.C. 35
Kinder, T. H., 1983, Donde Va: An oceanographic experiment near the Strait of Gibraltar, Proceedings of IAPSO/ONR/NEFKD Workshop on Straits, *Inst Phys Ocean*, Copenhagen, Denmark.
La Violette, P. E., 1981, Variations in the Frontal Structure of the Southeastern Grand Banks, NORDA Report 87, p 48.
La Violette, P. E., 1983, The Grand Banks Experiment: A Satellite-Aircraft-Ship Experiment to Explore the Ability of Specialized Radars to Define Ocean Fronts, *NORDA Report* 49, p 126.
La Violette, P. E., 1984, The advection of the submesoscale thermal features in the Alboran Sea gyre, *JPO*, v 14, no 2.
La Violette, P. E., 1986, Short-term measurements of the velocity of currents associated with the Alboran Sea Gyre during Donde Va, *JPO*, v 16, no 2, p 262–279.
La Violette, P. E., and P. L. Chabot, 1969, A method of eliminating cloud interference in satellite studies of sea surface temperature, Deep Sea Res, v 16, p 534–547.
La Violette, P. E., T. H. Kinder, R. Preller, and H. E. Hurlburt, 1982, Donde Va: a mesoscale flow dynamics experiment in the Straits of Gibraltar and Alboran Sea, *XXVIII Congress and Plenary Assembly of CEISM*, Cannes, France, December 2–11, 1982 (in proceedings).
La Violette, P. E. and H. Lacombe, Tidal-Induced pulses in the flow through the Strait of Gibraltar, *Oceanologica Acta* (in press).
McClain, E. P., W. G. Pichel, C. C. Walton, Z. Ahmad, and J. Sutton, 1977, Multichannel improvements to satellite-derived global sea surface temperatures, *Adv. Space Res.*, 2, 43–47.
Phillipe, M., and L. Harang, 1982, Surface temperature fronts in the Mediterranean from infrared imagery, in *Hydrodynamics of Semi-Enclosed Seas*, edited by J. C. J. Nihoul, pp. 91–128, Elsevier, New York.
Preller, R., and H. E. Hurlburt, 1982, A reduced gravity model of circulation in the Alboran Sea, in *Hydrodynamics of Semi-Enclosed Seas*, edited by J. C. J. Nihoul, pp. 75-89, Elsevier, New York.
Schwalb, A., 1977, Modified Revisons of the Improved TIROS Operations (ITOS D- G). National Environmental Satellite Service/NOAA, Washington, D.C., April *NOAA NESS Tech. Memo.* NESS 35, 48 p.
Smith, R. C. and W. H. Wilson, 1981, Ship and satellite bio-optical research in the California bight. *Oceanography from Space*, editted by J. F. R. Gower, Plenum Publishing Corporation, p. 281.

12
Image Processing In Optical Astronomy

Jean J Lorre
Jet Propulsion Laboratory
4800 Oak Grove Drive
Pasadena California USA

12.1. Introduction.

In the early nineteen seventies, a collaboration began between astronomers, notably Dr's Halton Arp and Jack Sulentic, at the Mt Wilsonand Palomar Observatories and image processing analysts working in the Image Processing Laboratory of the Jet Propulsion Laboratory. The problem presented was to attempt to apply some of the techniques developed at JPL for the unmanned planetary programme to certain difficult astronomical objects in order to determine their nature more precisely. This task resulted in the introduction of image processing techniques into astronomy. Most of this work involved the analysis of photographic plates and vidicon images. By the time CCDs became routinely available, a great many algorithms had been developed and the results of their application disseminated in the astronomical literature.

This Chapter will present a selection of the results of this on-going collaboration relating to a wide range of applications. In this manner, the author will attempt to provide a broad lever to arm the imaginations of those seeking tools for the future.

As you will realise while reading the following sections, the selection of the appropriate techniques from the image processing toolbox remains very much an art. Successful attempts at image enhancement often result from the exploitation of a 'weakness' in the image itself. This weakness is generally of three forms:

- Near symmetry of the object.
- A preferred directionality in the object.
- A distinction in spatial frequency between the object(s) and/or superimposed clutter.

In all cases, a price is paid for enhancing a property of the scene. The trick is to sacrifice a property that is not essential for interpretation.

12.2. Calibration.

It is of the utmost importance to calibrate a camera before attempting to acquire data with it. Astronomers have the choice of recording upon film or of recording electronically. In the latter case it is well worth the effort to perform a laboratory calibration of the detector using standard illumination sources. In the case of an electronic camera, a set of tests should be performed in a collimator exposing the instrument to flat fields of known exposure and to resolution and geometric targets. The tests should consist of:

1) Flat field exposures against an NBS standard through the filters to be used and bracketing the temperatures expected in the field. If this is not feasible then exposures should be acquired of the defocused dome.

2) Dark current integrations for the same durations as the flat field exposures.

3) Imaging through a collimator of a ruled-grid target.

4) Imaging an edge from which to determine the system modulation transfer function (MTF).

From these tests the following camera properties can be determined:

1) A response curve for each pixel at several temperatures.

2) A measure of the geometric distortion.

3) A measure of the MTF and its phase normal to the edge.

4) The coherent, periodic, and random noise as a function of exposure and temperature.

5) The temporal noise for each pixel, from which to determine a confidence interval for a radiance value.

6) A blemish file marking pixels which refuse to invert to a consistent exposure when applying the reciprocal response curve.

7) A file of saturation limits for each pixel.

Vidicons (Voyager Calibration Report) have several properties of which the user should be aware. They require a non-linear response curve for each pixel. It is usually most expeditious to create the response file by storing all the flat fields in a large file and to interpolate between them. This avoids assumptions about a model (whose coefficients require more storage space anyway). Radiometry is usually dependent upon the previous frames, which remain to some extent in the photocathode and leak out slowly. Vidicons are also subject to geometric distortion. A correction can be performed by moving the intersection locations of a calibrated grid target to positions consistent with the camera focal length. However it is important to beware that the geometric integrity of a vidicon is dependent upon the scene itself due to beam deflection from the cathode.

CCDs (Lorre, 1979, Galileo Calibration Report) are almost as complicated. The response of a CCD is linear at all but the lowest exposure regimes, thus storing only the gain for each pixel is satisfactory. Great care must be taken to ensure that detector saturation does not occur or else the gain computation will be meaningless. Consequently, it is often necessary to record the saturation value for each pixel in order to determine if a value is meaningful. Blemishes in CCDs can generate charge continuously, can block the transfer of rows or columns, and can leak into them. The MTF of CCDs is often misleading if they are aliased. Astronomers are usually saved from this phenomenon because the atmosphere bandlimits the signal before the detector can sample it. Aliasing has the effect of introducing uncertainty in point target locations and intensities, as well as confusing the interpretation of fine detail (Lorre *et al.*, 1980). Leakage from past frames also exists with CCDs, particularly from exposure in the red and the IR. An interesting radiometry problem with CCDs is interference fringes caused by thinning. These fringes can be corrected by applying the gain corrections computed from flat fields, provided the same filters are used. Line sources can only be corrected by using an equivalent line calibration source. It may be necessary to record sky flat-fields in order to record interference fringes caused by night-sky emission lines which would modulate the intensity of astronomical sources.

For all electronic detectors, the radiometric decalibration procedure consists of generating, by interpolation, bias and gain corrections. The steps are summarised below:

1) Interpolate a dark current (bias or shutter-inhibit) frame for the correct temperature and exposure as the image to be decalibrated.

2) Interpolate a flat field (for CCDs) or interpolate a non-linear calibration file (for vidicons) for the correct exposure and temperature.

3) Subtract the bias from the flat field(s).

4) Invert the response curve by dividing by the gain or by interpolating within the non-linear calibration file.

Photographic plates present very different problems. Radiometric calibration for these detectors is most often performed by scanning both the image area and a cluster of sensitometry spots placed on the plate through the same filter. The spots provide a single response curve for the plate. This is subsequently applied to every point on the plate. Within the same batch of plates of identical emulsion type the response is nearly identical. However, spatial variations across each plate are unique and non uniform. Methods have been developed for cases when no calibration exists (Sulentic, *et al.*, 1979) using models for the profiles of stars. In spite of these drawbacks emulsions have certain advantages over digital detectors, these are: large recording areas, absence of electronic interference, and non linearity which, if properly calibrated, results in great dynamic range.

Standard stars must be used to provide the connection to a reference through the attenuating effects of the atmosphere. This is true for any detector.

Figure 12.1a. Spatial variations in emulsion sensitivity apparent by severe contrast enhancement for the sky background. Slight variations of this sort are important when searching for faint surface brightness features with amplitudes measured in hundredth's of a percent of the total dynamic range. NGC1097, 45 minute exposure on 3aJ emulsion at the 4 meter telescope, Cerro Tololo by Dr H. Arp.

Figures 12.1a and 12.1b illustrate typical variations between type 3aJ emulsions for identical exposures of the same galaxy (NGC1097) after severe contrast enhancement. This sort of variation cannot be mitigated unless the structure of interest is small relative to the scale of the non-uniformity (Lorre, 1978).

As an example of calibration we present a procedure which was automated for the reduction of silicon vidicon spectra (Lorre et al., 1979). Figure 12.1 shows a mosaic of steps in the decalibration of the spectrum of a white dwarf. From the top down, the first four spectra are of: the distorted comparison spectrum, an incandescent spectrum, a dark current frame, and the spectrum to be decalibrated, respectively. The following steps define the automated procedure:

1) Locate and trace the stellar spectrum through each column of the image.

2) Model the stellar spectrum distortion with a polynomial.

Image Processing In Optical Astronomy 247

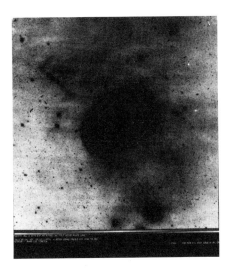

Figure 12.1b. Identical processing as figure 12.1a but in a different photographic plate.

3) Locate and trace each comparison line through each row.

4) Model each comparison line with a first order polynomial.

5) Determine the intersection points between the comparison line models and the stellar spectrum model.

6) Model the spectral distortion with a polynomial using the intersection points and the known wavelengths of the comparison lines.

7) Perform a geometric correction on the stellar spectrum correcting for both the spectral and off axis distortion. Sampling Theorem was used to avoid incorrect interpolation.

8) Perform the identical geometric correction on both the incandescent spectrum and the dark current. Rows 5–6 correspond to the points located on the comparison lines and stellar spectrum before and after distortion correction. Rows 7–8 correspond to the corrected stellar and incandescent spectra.

9) Remove the dark current from both the corrected stellar spectrum and the corrected incandescent spectrum.

10) Integrate the incandescent spectrum vertically to produce a profile.

11) Model the profile as a blackbody.

12) Generate a radiometric calibration file of gain corrections as a function of wavelength by dividing the model by the profile. Row 9 is a picture of the gain correction file.

13) Apply the gain corrections to the stellar spectrum. Row 10 is a

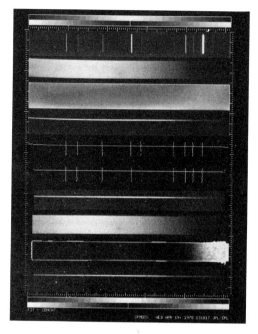

Figure 12.1c. Steps in decalibrating a spectrum obtained with a silicon vidicon. From the top down each image represents: 1. Comparison line spectrum. 2. Incandescent spectrum used as a black body model. 3. Dark current (response to no input signal). 4. Stellar spectrum to be decalibrated. 5. Locus of points traced by auto tracking algorithm. 6. Locus of points after geometric rectification. 7. Stellar spectrum after geometric rectification. 8. Incandescent spectrum after geometric rectification. 9. Image of computed gain coefficients used in radiometric correction. 10. Geometrically and radiometrically corrected stellar spectrum. Original data from Dr B. Oke, Mt Palomar Observatories.

picture of the corrected stellar spectrum.

14) Remove the sky by interpolating vertically from above and below the stellar spectrum.

15) Integrate the stellar spectrum vertically.

A great many other detectors exist, from electronographic cameras to microchannel plate front-ends to the above detectors. To a great extent they can all be calibrated by the same methods.

12.3. Enhancing Faint Surface Brightness Features.

One of the most challenging problems commonly encountered with astronomical images is the enhancement of subtle filamentary features. When photographic plates are used the variation in background between plates (or within a single plate) is often much larger than the amplitude of the features themselves. "Stacking" N images results in a reduction in the background, but only by the square root of N. Certainly the most powerful algorithm known for solving this problem is the median filter. This filter takes advantage of the fact that structure of interest is often of smaller spatial frequency than the background "clutter". It is also immune to Gibbs phenomenon, or the sensitivity which linear filters display to the presence of dazzling features like stellar point sources (which contain all spatial frequencies). The median level M of a neighbourhood with intensity histogram H(i) extending from 0 to N is

$$\sum_{i=0}^{M} H(i) = \sum_{i=M}^{N} H(i) \qquad (12.1)$$

To employ this algorithm a moving window is selected to cover an area at least three times the area of the largest object one wishes to retain. The window should have an orientation opposite to that of the object. In this way, the median filter will never respond to the object and will generate an image with a nearly faithful rendition of low spatial frequency structure which one does not want. Subtracting what one does not want from the original results in what one does want. An additional application of the filter restricted to a small window results in the rejection of all but the largest stellar sources. By treating a set of plates in this manner and then "stacking" them, one can arrive at a optimal gain in signal to noise (Lorre, 1978). Other non-linear methods have been used (Arp *et al.*, 1976), however linear filtering results in disasters. Great leverage can be gained over objects with symmetry or directionality to segregate different features much as an FFT does spatial frequencies. In the case of one-dimensional features recorded on two-dimensional media the filter can be used to advantage. This is spectacularly demonstrated in the case of spectra (Sulentic *et al.*, 1985).

A word of warning, non-linear filters of all sorts must be used with caution when they are the same size as the object(s) of interest. Corners can be removed and objects can be disfigured or vanish altogether in ways which are quite unexpected. When applied with caution this filter is very useful for enhancing spectra, modelling backgrounds, and segregating between classes of objects based upon shape.

There are several methods of computing the median other than the direct approaches of either re-computing the histogram or sorting the intensities. For small windows, one can make use of the last histogram of the adjacent overlapping window, by deleting the old column and adding the new one as the window shifts by one pixel. In addition, by keeping track of the side from which each old/new point comes from, relative to the old median, one can determine the direction to search in and the number of counts to accumulate before arriving at the next median.

Figure 12.2a. A single plate contrast enhanced to reveal faint surface brightness features. This processing corresponds to traditional enhancement methods. M82, 30 minute exposure on 103aO emulsion at the 200 inch telescope, Mt Palomar by Dr A. Sandage.

Figure 12.2a shows an image of M82 obtained by Dr. A. Sandage at the Palomar 200 inch telescope. Figure 12.2b shows the result of the addition of nine images, each median-filtered to remove low-frequency features, in order to improve the detectivity of thin filamentary structure present in the halo.

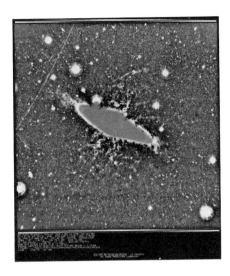

Figure 12.2b. Addition of nine images after processing with median filters to remove low frequency structure, followed by contrast enhancement. This processing is optimal for revealing faint filaments. Plates from Dr A. Sandage.

12.4. Automated Target Detection & Extraction.

Astronomical images often contain a large number of "target" images superimposed upon a relatively low-frequency background. This sort of data lends itself well to automated techniques for detecting the different targets and characterising them (Sebok, 1979). One technique in particular has been applied because it has the advantages of speed, modularity, and the ability to associate pixels from an object of arbitrary shape (Lorre *et al.*, 1979; Lorre , 1979; Lorre *et al.*, 1978). In outline form, the process is as follows:

1) Strip away a background model using a median filter of appropriate dimensions.

2) Select a threshold above which to consider pixels of interest.

3) Make a single pass through the image associating all pixels above the threshold. This can be performed by maintaining a buffer containing the identity number of each object and, by comparing new pixels with this running buffer, transferring the identity of the object (as well as other parameters) to the new pixels.

4) Maintain a catalogue of terminated objects, containing: area, intensity, moments, window dimensions, and orientation.

Once armed with a catalogue further analysis need not require revisiting the image. From the catalogue one can reject or accept candidate objects by virtue of their position in a classification space. Thus, one can screen for broad objects (low intensity-to-area ratios) or stellar objects (high intensity-to-area ratios) or objective prism spectra (width-to-height ratios).

A star-galaxy classifier was generated in this manner and presented images of Abell 6065, a cluster of galaxies superimposed upon a foreground of stars. From the catalogue, a classification space was constructed in which the target gaussian standard deviation was plotted versus the gaussian amplitude. This resulted in a segregation of targets into three classes: stars, galaxies, and unknown (too faint to determine a profile).

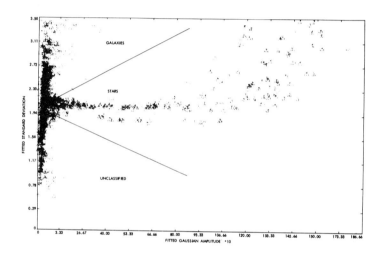

Figure 12.3a. Classification space for discriminating between stars and galaxies. The standard deviations and amplitudes of the radial profiles for each object are computed automatically. The space can be partitioned into three regions for object classification. AB-6065 48 inch Schmidt, Mt. Palomar.

Figure 12.3a shows this diagram. Figures 12.3b and 12.3c were generated by including only those objects considered to be galaxies and stars respectively. For a rigorous discussion of this problem see Sebok (1979).

For comparison, figure 12.3d illustrates the identification of suitable objective grating spectra (rejecting orders other than the first and images which were overexposed). The figure is taken from a small portion of a plate obtained by Dr. P. Osmer at CTIO (Osmer, 1982). These spectra

Image Processing In Optical Astronomy 253

Figure 12.3b. An image containing only the galaxies in AB-6065. Each image was removed from the scene based upon the classification of figure 12.4 and placed into an empty scene.

Figure 12.3c. An image containing only the stars superimposed over AB-6065.

were automatically integrated, subject to a search for Ly alpha emission indicating z=3 quasar candidates, and plotted. Object number 15 was a candidate. The emission line detection algorithm consisted of ratioing the spectrum candidate at each point in the sense: after median filtering/after linear filtering Narrow emission lines are unaffected by median filters.

254   J. J. Lorre

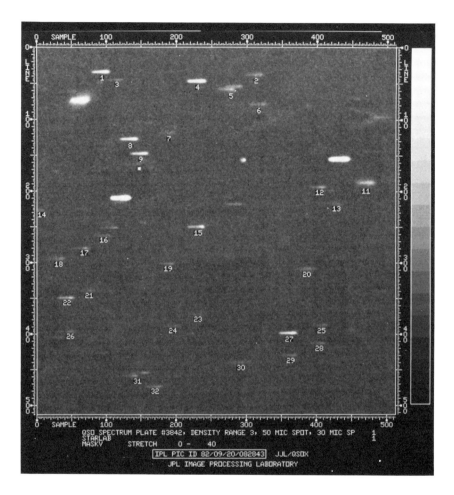

Figure 12.3d. A small part of a field containing objective grating spectra obtained in a search for high z quasars with Ly-alpha emission. All suitable candidate spectra have been automatically identified and catalogued before being subject to an algorithm which searched for emission lines. Object number 15 was selected as a candidate. CTIO Dr. P. Osmer.

Another area of great promise is the automated classification of galaxies by morphological type. Methods based upon structure in the Fourier Transform (Krakow *et al.*, 1982; Iye *et al.*, 1982; Sulentic *et al.*, 1984) hold the promise of extracting unique parameters from which different galaxies can be distinguished.

12.5. Geometric Transformations.

Geometric transformations of digital imagery are one of the most important tools available. They allow camera distortions to be removed (Voyager Calibration Report), different images to be registered digitally (Lorre, 1978), and symmetries to be exploited (Lorre, 1978, 1980).

This class of algorithm is set up as an inverse operator, i.e.: it asks of every pixel in the output image where it came from in the input image. Two polynomials are solved, one to transform from output to input images, and the other to interpolate between pixels in the input image since the mapped position in the corresponding pixel in the output image rarely coincides with an input pixel coordinate. Complexity arises over the method of accessing an input pixel when the input is too large to reside in memory.

An algorithm can be made quite general by allowing a large set of mapping points (tie-points) in both the input to output images. Several methods have been developed for determining which tie-points to use to describe the mapping of a particular pixel. One is to divide the space into triangles with tie-points at the vertices and to assign a polynomial transformation to all pixels within the triangle area. Another simpler approach is first to create a rectangular grid of intermediate tie-points. These are created by solving a low-order polynomial at each intersection by weighting the original points inversely by their distances. This new grid provides a simple partitioning of the image such that pixels are readily assigned to areas with their own transformations. In this way, highly non-linear transformations can be implemented without the requirement for a functional model to represent each conceivable application.

To illustrate such transformations figure 12.4a shows the polar coordinate mapping of a rectangular image of NGC1097 obtained by Dr. H. Arp at Cerro Tololo. The galaxy nucleus is stretched to fill the top of the frame, while the four corners map to the four peaks at the bottom. By computing the transformation about the nucleus we have unwound the spiral arms into straight lines canted at an angle.

As mentioned before, properties such as symmetry can provide a means of extracting information not otherwise readily available. We will take advantage of the symmetry of NGC1097 in its polar form to apply a radially-oriented median filter (a vertical filter in the polar space). Figure 12.4b shows the radial structure of the galaxy and the four jets emanating away from the nucleus (Lorre, 1978) without distractions due to foreground objects. The difference between the original image and the radial median shows the galaxy as it would appear in intermediate spatial frequencies devoid of the jets.

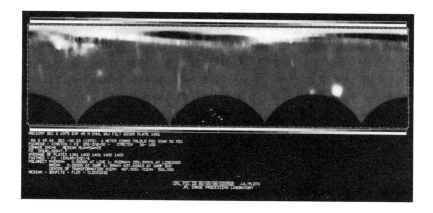

Figure 12.4a. A polar coordinate projection of NGC1097 with the origin at the nucleus. The nucleus maps to the top of the frame and the picture corners map to the bottom. This particular transformation provides cartesian oriented filtering algorithms access to the symmetric structure of the galaxy by restricting their operation to radial or azimuthal directions. 4m plate, Cerro Tololo, Dr. H. Arp.

12.6. Colour Space Transformations.

Thus far we have discussed operations using a single image or spectral band. Greater amounts of information are available when several images, each from a separate spectral region, are be combined. Quantitative maps of stellar type, age, or of physical properties such as electron temperature can be produced by combining bands according to astrophysical formulae (Dufour *et al.*, 1979; Dufour *et al.*, 1980; Kupferman, 1983). When relative radiances are not known but the corrections for linearity are known colour information can still be used to assist in spatially segregating the object into zones which are more readily discernible to the eye (Lorre, 1978).

N images, each in a different spectral band, create a colour space of N dimensions in which individual pixels map as vectors. If the space is quantised then the N-space becomes a hyper-histogram where clusters reveal the presence of classes of pixels with the same colour (spectral types, temperatures, chemical properties...). This is the classical Bayesian spectral classifier problem in which the object is to clump together in the colour space all clusters with the same spectral signature, thus revealing the image area belonging to each class. Techniques of this sort produce statistical data, but their application is often not optimal for the eye. This is because the classifier must be trained and one cannot train it without first recognising where the classes are. There is a solution, however.

Figure 12.4b. NGC1097 after a median filter was applied in the radial direction, followed by an inverse transformation back to cartesian coordinates. Only large features with radial orientation survive this operation, including the four radial jets discovered by Dr. Arp.

If one could move the clusters inhabiting the colour space apart, in such a way as to fill the space and do so in an orderly fashion, then the images which generate this new colour space would, when combined as a colour image, reveal strong colour discrimination. The reason why colour astronomical images of galaxies do not produce significant colour discrimination is because they are all highly correlated in intensity. A pixel which is dark in one image will tend to be dark in all the other spectral bands. This means that the region populated in the colour space will lie close to the diagonal, i.e.: the blue=green=red line (achromatic axis). We seek, therefore, to move clusters away from this location.

There are three types of known transformations which can generate such a colour mapping (Lorre *et al.*, 1981; Sulentic *et al.*, 1984), these are geometric, statistical, and repulsive.

Figure 12.4c. NGC1097 as it would appear with no large radially-oriented features. The nuclear region has been suppressed as well. This image was produced by subtracting figure 12.4b from the original.

Geometric colour transformation.

This class of transformations relies upon simple coordinate conversions to convert from the blue, green, red Cartesian system of the n-colour space to another set of coordinates in polar, cylindrical, or some other geometry. For example, the conversion to a polar space defined by first rotating the blue,green,red coordinates to x, y, z such that the achromatic axis (blue=green=red) axis coincides with the y axis followed by:

$$\text{Hue} = \tan^{-1}\left((z/x)^{1/2}\right) \qquad (12.2)$$

$$\text{Saturation} = \cos^{-1}\left((\frac{y}{x+y+z})^{1/2}\right) \qquad (12.3)$$

$$\text{Intensity} = x + y + z \qquad (12.4)$$

produces three new images with the properties of hue, saturation, and intensity. As in all these transformations the new images have special properties. In the above example the hue image contains the colour information and the saturation determines the purity of the colour. Typically the saturation will be very small for correlated scenes. One can achieve large colour discrimination by first multiplying the saturation by a large factor and then transforming back to the original blue,green,red Cartesian space (Lorre, 1976).

Statistical colour transformations.

This class of algorithm uses the data itself to determine the appropriate transformation. One of the most successful of these is called the Principal Component transformation. This is a linear rotation by matrix R upon the blue,green,red vectors for each pixel. R is computed from

$$R^t K R = K \qquad (12.5)$$

where K is the covariance matrix, t means transpose, and k is the matrix whose diagonals are the eigenvalues of the characteristic matrix of K. The valuable properties of the new images is that there is no longer any correlation between them, i.e.: the low saturation problem no longer exists.

Figure 12.5a shows a set of three images in different colours of the spiral galaxy M81 obtained by Dr. H. Arp at Mt Palomar. Figure 12.5b shows the three principal component images resulting from applying this transformation. These images are ordered in decreasing information content from the top down. Discrimination of colour classes is greatly simplified.

To achieve enhanced colour in the original blue,green,red sense one can perform a contrast enhancement upon each of the principal component images such that the histogram of each becomes gaussian, and then transform backwards to blue,green,red space using the inverse transformation of R. The gaussian histogram assures that the images remain uncorrelated, and the inverse transformation assures that the original colour sense is preserved. This transformation is not limited to three inputs.

Repulsive colour transformations.

This class of transformations treats each histogram element in the colour space as a repulsive body in a many body problem. The system is allowed to relax numerically until the colour space is filled. To date, attempts at implementing such an algorithm have been hampered by the memory and speed requirements imposed to maintain sufficient resolution. This is a maximum entropy solution to the colour discrimination problem and is the optimal solution.

260 J. J. Lorre

Figure 12.5a. Three images of M82 obtained through different filters. Direct superimposition through colour filters produces little dramatic colour because the images are highly correlated in intensity. Hale Observatories, 200 inch telescope, Mt. Palomar, obtained by Dr H. Arp.

12.7. Two Dimensional Histograms.

In the previous section on colour space transformations we referred to an N dimensional histogram as a classification space where pixels with common

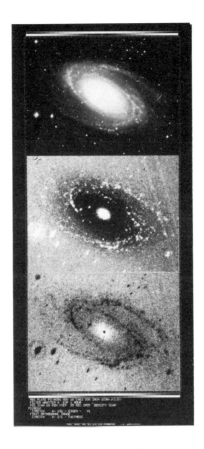

Figure 12.5b. The same three images as figure 12.5a after principal component transformation. The top image is a weighted sum and the others are linear combinations of the original images ordered by decreasing information content. These images are uncorrelated and represent good segregation of colours in the object. To produce a colour image with the correct blue, green, red sense these images should be contrast enhanced to produce Gaussian histograms and then the inverse transformation would be performed to revert from principal component space to blue, green, and red space.

signatures mapped to regions with specific meanings. Even without physical units such histogram spaces display interesting structure. When one can reduce the image data into the equivalent of two bands the histograms can be displayed as images. Images of NGC1097 obtained at Cerro Tololo by Dr. H. Arp in three spectral bands UV(103aO), Green(3aJ), and Red(127 and 098) were combined into two-dimensional histograms.

Figure 12.6a is an intensity-intensity diagram with UV on the vertical

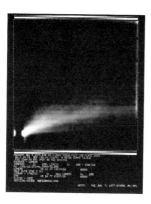 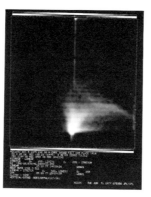 

Figure 12.6abc. (a) Two dimensional histograms generated by plotting colour pixels in a quantised intensity space. In this figure UV(103aO) is plotted vertically and red(098) is plotted horizontally. The sky maps to the bright cluster at lower left. These plots can be considered classification maps from which conventional Bayesian classifiers can determine class statistics for surface brightness. NGC1097, 4 meter plates, Cerro Tololo, by Dr. H. Arp. (b) Two dimensional histogram with UV(103aO) plotted vertically and red minus green (127-3aJ) plotted horizontally. This is surface brightness HR diagram. NGC1097. (c) Two dimensional histogram with UV minus green (103aO–3aJ) plotted vertically and red minus green (127–3aJ) plotted horizontally. This is a surface brightness colour-colour diagram. NGC 1097.

axis and red horizontal axis. The bright spot at lower left is the cluster due to the night sky background. Figure 12.6b is an intensity-colour diagram with UV on the vertical axis and red minus green horizontal axis. The vertical line is due to saturation in the red image. Figure 12.6c is a colour-colour diagram with blue minus green (vertical) and red minus green (horizontal).

Digital data placed into a calibrated magnitude system and converted to histograms could represent surface brightness HR and colour-colour diagrams (Dufour *et al.*, 1980; Dufour *et al.*, 1980).

12.8. Polarisation.

It is possible to compute maps of polarisation from images obtained through polarizers. This tends to be difficult because the computation involves differences between numbers which are nearly the same. Digital cameras provide the necessary precision.

In the following example three images of M20 were obtained at the Mt Wilson 100 inch reflector through a linear polarizer at positions of 0, 60, and 120 degrees. The recording medium used was film. Because no calibration was available the images were treated to set the sky level and the total variance in each image equal. This is not correct but it ensures that each image contributes the same information. If the intensities of the three plates can be represented as I1,I2,and I3, the polarisation, P, and the angle of the electric vector, A, can be computed (23):

$$K = (2/3)(I_1 + I_2 + I_3) \tag{12.6}$$
$$I = (8/9)((I_1 - I_2)^2 + (I_2 - I_3)^3 + (I_3 - I1)^2) \tag{12.7}$$
$$P = i/k \tag{12.8}$$
$$S = (4/3)^{1/2}(I_2 - I_3) \tag{12.9}$$
$$C = 2I_1 - K \tag{12.10}$$
$$2A = \tan^{-1}(S/C) \tag{12.11}$$
$$K = (2/3)(I1 + I2 + I3) \tag{12.12}$$

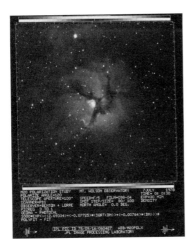
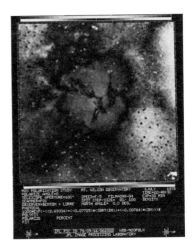

Figure 12.7ab. (left) Image of the reflection nebula M20. Mt Wilson Observatory, 100 inch telescope by W. Benton. (right) Polarisation map of M20 obtained by combining three images each obtained through a linear polarizer at 60 degree intervals. Polarisation of the dark sky is spurious because there is no signal.

Figure 12.7a shows one of the intensity images (they appear identical to the eye). Figure 12.7b shows the polarisation image. Areas away from the nebula which have no detectable intensity will produce large spurious polarisations. Figure 12.7c displays the polarisation vector, with the vector length proportional to the local polarisation.

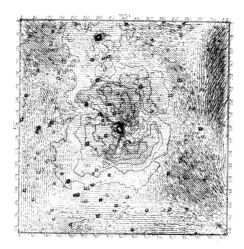

Figure 12.7c. Polarisation map of M20 with the electric vector direction indicated. The amount of polarisation is indicated by the vector length. Isophotes are superimposed on the image. These images are based upon a prescaling which sets the sky to the same value and the variances of each image the same. Such normalisations are common when dealing with uncalibrated regions and emulsions.

12.9. Resolution Restoration.

The problem of restoring the radiometric resolution in an image obtained through the atmosphere has a long history and, aside from speckle interferometry, the methods developed to solve it have not produced any great successes. This problem is due to the fact that the point spread function for long exposure stellar images is symmetric and nearly gaussian. The optical transfer function of a linear system is the complex FFT of the point spread function. This means that no phase distortion is present and thus the modulus of the optical transfer function can become zero. A zero, or nearly zero, transfer function means that no information is passed beyond a certain frequency and there is nothing to restore. Really successful models will have to depend upon *a priori* knowledge of the image or the physics which it represents.

A number of interesting algorithms has been developed however (24–26) some of which will be described here. If we define P as the point spread function and OTF as it's FFT then the simplest model for image restoration is the Wiener model

$$\text{correction} = \frac{\text{OFT}^\times}{|\text{OFT}|^2 + 1/\text{SN}^2} \qquad (12.13)$$

where SN is the signal to noise ratio and '×' means complex conjugate. One can achieve good cosmetic restoration by allowing SN to vary in the image by assigning it to the local standard deviation. This requires one to abandon the use of FFT's however.

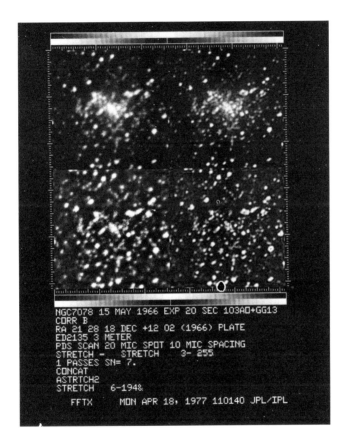

Figure 12.8. Linear convolution restoration for atmospheric blurring using Wiener filtering. The original image is at upper left. Below it is the original image with low spatial frequencies removed. To the right of each is its restoration. It is often advantageous to incorporate other filters with the restoration filter to aid the visual interpretation. Radiometric resolution gains are limited in practice to around 20%.

Figure 12.8 shows a cluster of stars NGC7078 in which the original image is at the upper left, with a high-pass version below it. The high-pass is intended to remove the dynamic range of the cluster. Each image on the left-hand side has been restored with the Wiener model and the result is displayed on the right-hand side. The radiometric resolution has been improved by about 20%. It is clear from this figure that operations other than simple restoration can assist in the interpretation of detail. Wiener models, as well as many other similar models, assume no *a priori* knowledge of the object and this is rarely the case.

Iterative methods exist which are simple, moderately efficient, and do not require the solution of the inverse problem. If we define M_i as the i(th) estimate of the true scene, I as the blurred image, and @ as convolution then a maximum entropy model predicts the next estimate M_{i+1} from:

$$X = (I - M_i @ P) @ P \tag{12.14}$$
$$L = (\log(d) + 1)/\max(\mathbf{X}) \tag{12.15}$$
$$M_{i+1} = M_i + \exp(LX - 1) \tag{12.16}$$

$\max(\mathbf{X})$ is the largest absolute value of all the pixels in X, and L is a constant around 10 which regulates the rate of convergence.

Another scheme performs estimates of the amount of P to add at a grid of points, in order to convert the latest convolved estimate into the blurred image. In this case:

$$X = I - M_i @ P \tag{12.17}$$
$$L_i = \text{sign}(X_i) \max(\mathbf{X})_j, \quad \text{if } M_{ij} > 0 \tag{12.18}$$
$$M_{i+1,j} = M_{ij} + L_j/P_0 \tag{12.19}$$
$$\text{if } M_{i+1,j} < 0 \text{ then } M_{i+1,j} = 0 \tag{12.20}$$

J is the pixel and P_0 is the amplitude of P at the centre after the integral of P has been normalized to unity.

12.10. Rotation Curve Determination.

One of the applications we have found useful for correlation is in determining rotation curves. Spectra of extended objects in rotation display a doppler tilting of the spectrum from which the rotation curve of the object can be determined. If the spectrum is faint or the lines of poor quality it can be difficult to determine an accurate rotation rate. One method is to select a strip down the centre of the spectrum in the dispersion direction, compress it into a single vector, and then correlate it with strips taken at regular intervals at different positions in the cross-dispersion direction. This process integrates all of the information in the spectrum obtained at a particular source position greatly improving the signal to noise.

Figure 12.9a shows a spectrum of Saturn obtained by Dr M. Belton at KPNO. The tilt in the spectral lines is apparent. Figure 12.9b shows an

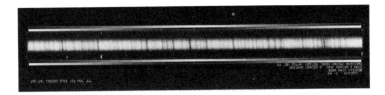

Figure 12.9a. Spectrum of Saturn with the slit placed along the equator. Spectral lines are Doppler tilted indicating the rotation rate. KPNO by Dr. M. Belton.

enlarged region about the correlation centre. Each maximum has been set to black as a visual aid. In this case, the rotation rate could be determined automatically.

12.11. Conclusion.

The field of image processing has become an indispensable aid to the modern astronomer with his reliance upon digital detectors. Space based detectors return all of their data in digital form whether they look up or down. Hopefully the examples provided in this book will inspire others to explore avenues not yet discovered.

Acknowledgements.

I am indebted to Dr's Halton Arp, Jack Sulentic, and Bruce Goldberg for their persistence in assisting to develop this new field at JPL.

References.

Arp, H. et. al., 1976, Image processing of galaxy photographs, *Astrophys. J.*, 210, 58

Dufour, R. et. al., 1979, Picture processing analysis of the optical structure of NGC5128 (Centaurus A), *Astron. J.*, 84,284

Dufour, R. et. al., 1980, Anatomy of a spiral galaxy, *Sky & Telescope*, July, 23

Fienup, J.R., 1978, Reconstruction of an object from the modulus of it's fourier transform, *Optics Letters*, Vol3, No1

Galileo Calibration Report. (not yet published)

Gull, S.F., 1978, Image reconstruction from incomplete and noisy data, *Nature*, 272, No5655, 686

Iye. M. et. al., 1982, Spectral analysis of the asymmetric spiral pattern of NGC4254, *Astrophys.J.*, 256, 103

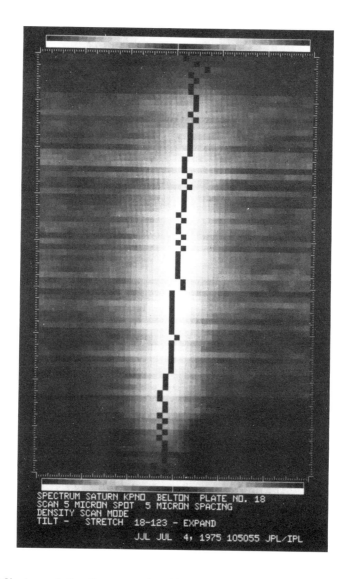

Figure 12.9b. A correlation diagram produced by correlating a strip obtained from the centre of the spectrum with strips at different vertical offsets. The correlation peak has been marked black. Such diagrams can be used to determine a rotation curve when the individual spectral lines are too faint to follow.

Krakow, W. *et. al.*, 1982, Analysis of the power spectra of galactic images and spiral structures, *Astron.J.*, 87,203

Kupferman, P.N., 1983, Two dimensional photometry of planetary nebulae, *Astrophys.J.*, 266, 689

Lorre, J.J., 1977, Application of digital image processing techniques to astronomical imagery, *JPL Publication*, 78–17 .

Lorre, J.J., 1978, L99 enhancement of the jets in NGC1097, *Astrophys.J.*, 222, No.3, Part 2.

Lorre, J.J., 1979, Application of digital image processing techniques to astronomical imagery, *JPL Publication* 79–109 .

Lorre, J.J., 1980, Application of digital image processing techniques to astronomical imagery, *JPL Publication* 81–8.

Lorre, J.J. *et. al.*, 1979, Recent developments at JPL in the application of image processing to astronomy, *SPIE* 172,394

Lorre, J.J. *et. al.*, 1980, Artefacts in digital images, *SPIE* 264,123

Lorre, J.J. *et. al.*, 1981, Multispectral enhancements for astronomical images, *J.Brit.Interplanetary* Soc.34, 139

Lorre, J.J. *et. al.*, 1984, Image restoration of high resolution observations of the M87 jet, *Astron.Astro.*, 130, 167

Lorre, J.J. *et. al.*, 1978, Application of digital image processing techniques to astronomical imagery, JPL Publication 78–91.

Osmer, P., 1982, Evidence for a decrease in the space density of quasars at $z \geq 3.5$, *Astrophys.J.*, 253,28

Redman, R.O. *et. al.*, 1986, The dust and gas surrounding Lick H-alpha 101, *Astrophys.J.*, 303,300

Sebok, W.L., 1979, Optimal classification of images into stars or galaxies—a Bayesian approach, *Astron.J.*, 84, 1526

Sulentic J. *et. al.*, 1983, Analysis of optical imagery for Seyferts Sextet and VV172, *Astron.Astrophys*, 120, 36

Sulentic, J.W. *et. al.*, 1979, Some properties of the knots in the M87 jet, *Astrophys.J.*, 233, 44

Sulentic, J.W. *et. al.*, 1984, The magic of image processing, *Sky & Telescope*, May, 407

Sulentic, J.W. *et. al.*, 1985, Optimal enhancement of features in digital spectra, *Astron.J.*, 90, 522

Voyager Calibration Report, Voyager imaging science subsystem calibration report (for more information: 618–802, july1978, Benesh M.)

Index

A-buffer 29.
Absolute orientation parameters 31.
Achromatic axis 258.
Airborne Expendable Bathythermographs (AXBTs) 234.
Albedo 171.
Alpha model 197.
Angstrom exponent 229.
Anisotropic scale invariance 183.
Anti-aliasing 99.
Apparent Thermal Inertia (ATI) 162.
Apple Macintosh 10.
Applications software 77.
ASCII 39, 42, 45, 129, 130, 131.
Arithmetic Logic Units (ALUs) 5.
Arrays 88.
    boolean arrays 86.
Atmospheric Correction 157.
Atmospheric and Oceanographic Information Processing System (AOIPS) 60.
Auto-correlation 79.
Automated stereo matching 24.
AVHRR 220, 224.
AXBTs 235.

BBC microcomputer 127.
BISHOP 61.
Baud rate 111.
Bayesian spectral classifier 256.
Beta model 197.
Beta model 199.
Bi-directional Reflectance Distribution Function 21, 29.
Bit-mapped technology 15.
Blackbody 247.

CCDs 243.
CD-ROM 9, 15.
CZCS 229, 233.
Canadian meteorological network 191.
Canadian surface meteorological network 207.
Cascade models 182.
Cascade processes 195.
Cellular Logic Image Processor (CLIP4) 6.
Colour Space Transformations 256.
Colour framestores 109.
Colour space transform
    Geometric colour transformation 258.
    statistical colour transformation 259.
Command language 80.
Command procedures 48.
Computer Hardware sub reduction in cost 105.
Computer compatable tapes (CCTs) 109, 115, 224.
Conditional array processing 73.
Connected component labelling 3.
Constant Altitude Z Log Range maps (CAZLORs) 201.
Contrast enhancement 71.
Convolution 99.
    Butterworth Filter 99.
    high pass filter 99.
    low pass filter 99.
Coriolis force 196.

DEC 106.
DEC PDP11 105.

DECNet 53.
DEM 169.
DIAD 106.
DIAD 113.
DIPIX 106.
Data compression 109, 125, 159.
  Run length coding 159.
Data structure 84, 86.
  data format 84.
  data types 84–85.
  scalar 84.
Data transfer 130.
Deanza 80.
Delauney triangulation 24.
Device independence 76.
Diffuse attenuation coefficient 218.
Digital Elevation Models (DEMs) 21.
Display device 76.
Distributed processing 16.
Dyadic operations 84–85.

ERS-1 9.
ESA 9, 120.
ESOC (European Space Operations Center) 157.
Edgels 3.
Effective spectral radiance emittance 216.
Entropy 259.
Error diffusion algorithms 10.
Ethernet 15.
Extreme variability 179.

FORTRAN 127.
Fileservers 118.
First optical attenuation length 218.
Floppy disks 118.
Fourier transform 79, 93, 98.
Fractal Sums of Pulses (FSP) 189.
Fractal dimension 179, 186, 206.
Fractals 179.
Fresnel reflection 218, 220.
Function 83.

GEMS of Cambridge 106.
GOES 220.
Gaia 9.
Gargantini list 138, 141.

General Meteorology Package (GEMPAK) 60.
Generalised Scale Invariance 179, 196, 205.
Geographical Information System (GIS) 112.
Geometric correction
  navigation 155.
  registration 154.
Geosynchronous orbits 220.
Gibbs phenomenon 249.
Goddard Space Flight Center 40, 59, 60.
Gradient Operations 100.
  Roberts operator 100.
  Sobel operator 100.
Graphics tablet 65–66, 76.
Gravimetric data 167.

HCMM 153, 161–162, 164.
HSI transformation 259.
Hard-copy Output Peripherals +2 113.
Hewlett-Packard 106.
Hierarchical data structures 136.
Host computer 55, 127.
Human Computer Interface (HCIs) 4, 10, 12, 39, 41, 58, 65–78.
Hyperbolic intermittency 196.
Hyperbolic intermittency parameter 184.

I^2S 5.
IAX 12, 79–102.
IBM 10, 12, 79, 83, 91, 124.
  IBM 5080 display 80.
IBM PC 118.
IDP 3000 125.
IPIPS 6.
Image processing languages 11.
Image refresh memory 76.
Image syntax 135.
Imperial College 5.
Inmos transputers 7.
Input buffer 67–68.
Instantaneous field of view (IFOV) 111.
Integral structure function 181, 187, 199, 207.

Intel 86 106.
Interactive Menu Interface (IMI) 65–78.
Interactive Planetary Image Processing System (IPIPS) 5.
Interactive Research Imaging System (IRIS) 60.
Interlace flicker 67.
Intermittency exponent α 189.
International Imaging Systems 106.

Jet Propulsion Laboratory (JPL) 1, 22, 60, 232, 243.

Kirchhoff's law 216.

LANDSAT 9, 105–106, 109, 115, 123–124, 153, 157, 162, 164, 169.
   MSS 109–110, 115, 158, 161–162.
   TM 109–110, 115.
LANDSAT Assessment System (LAS) 60.
LS10 131.
Lambertian reflectance 26.
Laser-Scan Laboratories 24, 33.
Laser-disks 9.
Lawrence Berkeley Laboratory 41.
Levy flight 193.
Lineaments 164.
Literal 84, 86.
Local area network LAN 115.
   Ethernet 115.
LogEtronics 106.
Look-Up Table (LUT) 2, 51–52, 68, 76, 100, 126, 128–129.

MCSST 227.
METEOSAT 95, 153–157, 171, 220.
Machine code subroutines 127.
Map projections
   Lambert conformal conic projection 158.
   Universal Transverse Mercator 158.
Maximum entropy model 266.
Mediterranean Sea 227, 233.
Mercator projection 224.
Meteosat 5.
Microcomputer Chip Characteristics 107.
Microcomputer clock speed 106.
Microcomputers 105.
Microsoft 13.
Microsoft BASIC 127.
Mie scattering 219.
Models of image formation 23.
Modulation transfer function 244.
Monadic operator 92.
Monte Carlo Ray-tracing 27, 29.
Motorola 68000 106.
Mouse 65, 75.
Mt. Wilson Observatory 243.
Multichannel Sea Surface Temperature (MCSST) 226.
Multichannel sea surface temperature 227.
Multimission Image Processing Laboratory (MIPL) 60.
Multiple Instruction Multiple Data (MIMD) 7.

NASA 1, 9, 40, 118, 124, 130, 215.
NASA Goddard 77.
NASA Space Science Data Center 61.
NASA/Ames Research Laboratory 61.
NOAA 36, 120, 156–157, 161, 213, 215, 220, 230.
NOAA AVHRR 93, 153–154, 155, 171.
NOAA AVHRR-IR 226.
NOAA National Environmental Satellite Data Information Service (NESDIS) 226.
NPA Ltd 118.
National Remote Sensing Centre 124, 131.
Nature Conservancy Council 131.
NeWS 13.
Nimbus 213, 220, 230.
Nimbus-7 CZCS 232.

Nimbus-7 Coastal Zone Colour Scanner (CZCS) 221.
Noah effect 208.
Noise removal 95.

Oceanographic Applications 213.
Operator 83.
Operator Priority 87.
Operators 86.
    boolean operator 86.
    concatenation operator 86.
Optical Disks 119.
Optical transfer function 264.

PCIPS 81.
PDP11 106.
PRIME 106.
PRTV 82.
Palomar Observatories 243.
Parallel Processing 5.
Parallel interface sub centronics 111.
Parallel processing 9.
Parameter array 71.
    Status flags 71.
    Status flags 72.
    array index 71.
    conditional array 72.
    numeric file input/output parameters 72.
    numeric input parameters 72.
    refresh memory array 72.
Peripherals 112.
    digitising tablet 112.
    video digitisers 112.
Phytoplankton concentration 230.
Pilot Climate Data System (PCDS) 60.
Planck's law 215.
Pluto micrographics system 126.
Point spread function 264.
Polar Stereographic projection 224.
Polarisation 262.
Preston Polytechnic 131.
Principal Component transformation 259.
Problems of magnetic tape 9.
Process definition 71, 74.

Process file 70, 74.
    interactive mode 71.
    non-interactive mode 70.
    semi-interactive mode 70.
Programming language
    ALGOL-60 81, 90.
    APL 81.
    123, 127.
    C 12, 58.
    FORTRAN 56, 76, 81.
    GOP 81.
    LISP 82.
    PASCAL 81.
    PICAP 81.
    PL/1 80-81, 90.
    PROLOG 82.
    array oriented language 81.

Quadtree storage
    bottom-up-quadtree 136.
    top-down-quadtree 136.
Quadtrees 136.

RAM 4, 23, 105.
RISC processor 3.
RS232 127.
RSRE Malvern 7.
Radar 178.
Radiometric calibration 157, 226.
Radiometric de-calibration 2.
Radiosity 27.
Ramtek 9400 80.
Random access memory (RAM) 111.
Raster systems 75.
Rayleigh scattering 219, 229.
Refresh memories 51–52.
Rendering 26.
Repulsive colour transformations 259.
Root mean square 155.
Rotation Curve Determination 266.
Run-length encoding 141.

SEASAT 220.
SIMD 7.
SPIDER 12.
SPOT 36, 105.
SPOT IMAGE 120.

HRV 110, 115.
SUSIE 81.
Scalar 86.
Scale invariance 182, 183.
Serial interface
 RS232 111.
Signal processing 80.
Signal-to-noise ratio 227.
Single Instruction Multiple Data (SIMD) 6.
Sobel operator 164.
Software 73.
Software Portability 56.
Soil surface heat flux 172.
Sonobuoy drifters 234.
Space Shuttle 218, 235, 237.
Space Station 61.
Spatial resolution 110.
Spectral emissivity 216.
Spectral radiant emittance 215.
Star-galaxy classifier 252.
Status flag 73.
Stellar spectrum 246.
Structure functions 184.
Sub-sampling 111.
Sun 10, 28.
Sun Microsystems 108, 115.
Sun TACC-1 7.
Sun-NFS 15.
Sun-RPC 15.
Sun-synchronous orbits 220.
Surface shading models 21, 26.
Syntax 68, 80.
Synthetic Aperture Radar (SAR) 237.
Synthetic reflectance image 169.

Transportable Applications Executive (TAE) 11, 13, 77.
 History of TAE 40.
 TAE Classic 42.
 TAE Plus 42, 49, 55, 57.
TCP/IP 53.
Teal Ruby Experiment 61.
Terrain visualisation 22.
Tesseral addressing 136.
Tesseral arithmetic 141.
Tesseral raster storage 139.

Thematic Mapper 60.
Thermal inertia 174.
Trackerball 65.
Trackerball 75.
Transport and Road Research Laboratory 124.
Transportable Applications Environment (TAE) 39.
Transputer 36.
Transputers 16.
Triangulated Irregular Network (TIN) 24.
Turnkey 65, 107.

UNIX 58–59.
University College London 5.
Unix 12.

VAX 5, 58, 68–69, 106.
VAX/VMS 12, 59.
VICAR 12, 61, 81.
VLSI 22.
Variable 84–85.
 complex variable 88.
 non-complex variable 88.
 pseudo-variable 90, 100.
 pseudo-variable assignment 90.
Vector 86.
Vector format 112.
Vectors 79.
Vicom 106.
Virtual operating system 41.
Voyager 1.

WIMP 4, 13.
WMCE 234.
WMCE 237.
WYSIWYG 71.
Western Mediterranean Circulation Experiment (WMCE) 233.
Western and Eastern Alboran Gyres 234.
Wide Area Networks 16.
Widening rule 85.
Wiener filtering 264.
Winchester disk 117.
Write Once Read Many (WORM) 9.

X11 13.
Xerox Palo Alto Research Center 13.

Z-buffer 28–29.

Z80 microprocessor 127.